THE PRINCIPLES OF
OBJECTIVE TESTING IN MATHEMATICS

THE PRINCIPLES OF OBJECTIVE TESTING SERIES
General Editor: Douglas M. McIntosh
C.B.E., LL.D., M.A., B.SC., M.ED., PH.D., F.R.S.E., F.E.I.S.
Principal, Moray House College of Education

THE PRINCIPLES OF

Objective Testing in Mathematics

W. G. FRASER, BSc., A.F.I.M.A
Moray House College of Education

J. N. GILLAM, M.A.
Scottish Certificate of Education Examination Board

HEINEMANN EDUCATIONAL BOOKS
LONDON AND EDINBURGH

Heinemann Educational Books Ltd

LONDON EDINBURGH MELBOURNE TORONTO
JOHANNESBURG AUCKLAND SINGAPORE KUALA LUMPUR
HONG KONG IBADAN NAIROBI NEW DELHI

ISBN 0 435 50330 8
First published 1972

A/510.76

Published by Heinemann Educational Books Ltd
48 Charles Street, London, WIX 8AH
Printed in Great Britain by
Butler & Tanner Ltd, Frome and London

Preface

Some form of assessment is necessary if teaching is to be effective. The objectives of any lesson or series of lessons must be clearly defined, learning experiences planned, and assessment made of the success with which the pupils have attained the stated objectives. Assessment therefore involves both an attempt to measure the ability of the pupil to benefit from the instruction and also the effectiveness of the teaching.

Tests and examinations may vary from the formal type of external examination to the more simply compiled classroom tests. The former must be carefully constructed as their results are often a major factor in determining pupils' careers; the latter do not require the same detailed care and attention, as they normally are attempting to confirm whether pupils have mastered part of a course of study and the results are interpreted in the light of the teachers' knowledge of the pupils' work in the classroom.

There are many criticisms of formal examinations, particularly when the results do not take into account teachers' estimates of pupils' ability or achievement. For this reason it is often suggested that a continuous record of the pupil should be kept in the school and in this way a much more reliable assessment can be given about the pupil. Objective tests can be used most effectively by the teacher in building up records of achievement of pupils throughout the school. Examining bodies are also using objective testing to a greater extent: as teachers must prepare their pupils for these examinations, it is essential that they understand how they are constructed and how the pupils can best be prepared for them.

1972

W. G. F.
J. N. G.

Contents

1. *The Purpose of Examinations*

Assessment is an essential element in education. The teacher must make some measurement of the extent to which the learning experiences of pupils have enabled them to achieve the stated objectives of the course of study. For example, if pupils have been taught Pythagoras' theorem, the teacher must check whether they understand the theorem and can apply it to new situations.

Several methods are open to the teacher to find out the extent to which pupils are successful in achieving the objectives of the course which they are studying. Exercises in the classroom, oral questioning, and written examinations are all part of normal classroom practice. Each has special features which limit its efficiency as a measuring instrument. The written examination has been shown by experimental techniques to yield marks which may be inconsistent: even the same examiner will award the same scripts widely different marks on separate occasions. The development of objective testing has helped to remove this particular weakness in examinations.

Consistency of marking by itself, however, would probably not have been sufficient to give objective testing the general support it now receives. Its ability to test for specific educational objectives has provided teachers with a much more accurate means of establishing pupils' achievement. Understanding and application can be assessed in much greater detail, and a clearly defined specification for a test ensures that a representative sample of the syllabus has been tested. Objective tests have also the incidental advantage that they can be very rapidly marked: a process which employed one hundred markers over four weeks can be replaced by one which uses a machine for a few hours.

Examinations, particularly external examinations, have been criticized as having a bad effect on education. Such criticisms have generally been justified only when the examinations used were of poor quality, and results were assumed to have an unwarranted

1

accuracy. Throughout an educational system there must be standards which pupils are expected to achieve and the attainment of which justifies proceeding to a higher stage in the study of a subject: the pupil who cannot master elementary algebra is unlikely to be successful in calculus. Examinations also play a vital part in the evolution of education: innovation will be based on opinion unless evaluation techniques provide the evidence to justify the proposed changes. The question is not whether there should be examinations, or tests, but how can they be improved in their indispensable role as part of the education process.

PURPOSE OF EXAMINATIONS

When an examination is being constructed or selected for use, the first consideration must be a clear understanding of the purpose of the examination. External examinations set to a large number of pupils from different schools by competent examiners, provide evidence that a particular level of knowledge and abilities has been acquired. Such examination results are also accepted by prospective employers as indicating desirable qualities of character and in particular testifying to general all-round ability which will enable the holder to undertake work of a particular character. University authorities generally play a part in determining the nature of the examinations which are acceptable for determining entrance qualifications.

Examinations or tests can serve six main purposes: (a) the measurement of achievement, (b) selection, (c) prognosis, (d) diagnosis, (e) motivation, (f) a teaching instrument.

(a) *Achievement*

To measure the achievement of a pupil after completing a prescribed course of study is perhaps the simplest and most common purpose of an examination, whether it be external or teacher-made. To construct such an examination the objectives of the course must be clearly defined in terms of human behaviour. There is no way of seeing into a pupil's mind to determine what he knows; it is only possible to determine this by some aspects of his behaviour or performance and this can be verbal or non-verbal. To find out whether a pupil can solve quadratic equations, the teacher must set a test consisting of different types of such equations which the pupil has to solve. Ability to solve equations, however, does not indicate the ability to apply quadratic equations to problem solving.

(b) *Selection*

Selection is inevitable when there is a limited number of places, whether it be to the Administrative Grade of the Civil Service or to a faculty of a university, in which there are a great many more applications for admission than there are places available. Such examinations are not of the pass/fail category since what is required is a rank order of candidates. It is generally assumed that the examination can distinguish between candidates whose marks are not widely scattered.

(c) *Prognosis*

Examinations are used to predict what the examinee is likely to be able to achieve in the future. A high level of achievement in mathematics in later years of the secondary school is a good indication that the pupil is intellectually capable of being a good student of the subject at university. On the other hand, although high marks in examinations in the early years of secondary education may also be indicative of high intellectual ability, the pupil may show little interest or inclination to study the subject in later years.

Examinations used for selection are assumed to have a predictive value. The predictive value of a test or examination can only be determined by comparing a pupil's performance on the test with attainment of the pupil in the course for which the test was used as a predictor. Within a school, teachers should be able to judge the effectiveness of their prediction. There are relatively few studies which establish accurately the value of school examinations in predicting university success. Again, teachers themselves can judge, over long experience, the type of pupil who is likely to make a success of university education. Test results in the primary school do not predict for all pupils the success which they are likely to achieve in the wide variety of subjects in the secondary school.

In the school, any form of 'setting', 'streaming' or grouping of classes by means of a test or group of tests is, in fact, using them for the purpose of prognosis.

(d) *Diagnosis*

Diagnostic tests are constructed not to assess achievement in a subject but to reveal the weaknesses of pupils in a particular section of the work, although mistakes made in attainment tests can be indicative of difficulties, thus also providing diagnostic information. Such tests are used widely in primary schools, particularly in English and arithmetic, and the diagnostic function of the school test can also be an important aspect of secondary level.

An examination can be of diagnostic value in answering the question, 'Has the teaching of a part of the work been effective?', the point being that a poor response may be due to inadequate teaching or choice of subject material. To the teacher this feedback information about the effectiveness of his teaching is important.

The end result of proper treatment of the findings of a diagnostic test can be course improvement, teaching improvement and pupil performance improvement.

In the early years of the secondary school, tests designed to serve the purpose of diagnosis are of value in ensuring that pupils are given a good grounding in subjects which are new to them. The onus of this type of work must lie to a large extent with the teacher, and the class examination must be constructed with the purpose of diagnosis in mind. Internal examinations tend to measure attainment but there should also be an emphasis on the diagnostic function of examinations.

(e) *Motivation*

Motivation is acknowledged as one of the powerful factors in the learning process. Whether the tests be external or teacher-made, they can be an incentive and spur to both pupils and teacher. Particularly in the secondary school and at university, they provide an objective and stimulus to schools and individual students. Incidence of periodic tests and examinations compels pupils to organize their work-habits more systematically and effectively than they would if they were free from the commitment of such demands. Examinations also bring benefits for teachers in that the work of the syllabus has to be planned systematically, taught thoroughly, and the individual teacher has a standard against which to measure his class.

(f) *Instrument of Teaching*

Examinations can help the teacher, as well as the pupils, in the learning process. To make education effective, teaching objectives must be defined, learning experiences be provided to achieve these objectives and examinations be part of the process that determines the extent to which the objectives are being attained. Unless the teacher in the class designs an examination with particular objectives of the course in mind, its use as a teaching instrument will not be fully exploited. Such tests require to be given at various stages of the course, not only at the end of a term but also through the medium of short tests set regularly in the classroom.

The use of an examination as an instrument of teaching in the sense of compelling the teacher to think about the objectives being tested when he is constructing the test, and in the light of the results reviewing the learning experiences given to the pupils in attempting to attain these objectives, must lead to more effective teaching and consequent improvement in the education process.

Multipurpose nature of examinations

In the ideal situation an examination would be constructed for the one well-defined purpose it is required to serve and its results would be limited to that purpose. In general, examinations are multipurpose. A test constructed to measure one function, for example to certify that a pupil has completed satisfactorily a course of study, will not necessarily be successful in predicting future performance, providing incentives for pupils and teachers, discriminating between candidates, or acting as a guide to effective teaching.

At present, examinations are made to serve too many purposes and, especially with external examinations, it was only when the original purposes were obscured by the passage of time and by changing conditions that a particular examination became a target for criticism. For example, certificates at the end of secondary schooling were awarded on the basis of examinations, to mark the successful completion of a course of study. Later they were extensively used as admission qualifications to higher education and a variety of careers: there is little evidence to establish that they can achieve these different objectives with equal success.

In school examinations teachers have the opportunity to avoid this confusion of purposes by ensuring that the examination or test is constructed with a specific purpose or purposes in mind. To do this they must acquire a deeper insight into the principles and techniques of testing.

Critics of examinations see them as curbing the development of interest, controlling the curriculum and teaching methods, and failing to serve adequately the purposes for which they are employed. National external examinations will continue to some extent to control and dictate how and what the teacher will teach: examiners and teachers must accept this situation but they must ensure that examinations are constructed in such a way that they will influence teaching to achieve the ends which are accepted as being educationally desirable.

EXAMINATION AND TEST RESULTS

Generally, examination or test results are expressed as a total of the marks for each question although a pupil's total performance may not be significant. For example, in a test used for diagnosis it is the pupil's failures on specific questions which have to be studied, and similarly where the test is used as a teaching instrument, the pupil's success in individual questions may point to the effectiveness of the teacher in specific areas of the course of study.

The results of tests can be expressed by:

1. numerical marks,
2. categoric marks,
3. rank orders.

1. *Numerical Marks*

(a) *Raw Score*

The total performance of a pupil is generally signified by a numerical mark. In an objective test, as one mark is generally awarded to each correct answer, the total number of correct answers is the total score. The sum of marks for each question is termed a raw score and by itself is of little significance, even when the total number of questions is also given: the average marks of a class at least must be known in order to indicate the level of difficulty of the test.

(b) *Standard Score*

A mark has significance only when it is related to the marks of a fairly large number of pupils of the same age or at the same stage in a course of study. 60 may indicate a good performance or a poor one, depending upon the achievement of other pupils who have sat the test. For example, if the average mark is 45 then 60 is an above-average score, but if it is 70 then the same mark represents a below-average performance.

Even knowledge of the average mark is insufficient to give meaning to a mark since, in some cases, the scores will be spread over a wide range whereas in others the range may be a narrow one: mathematics tests generally give a wide scale of marks, but in subjects such as art the range of marks is restricted. When scores from different tests are added, the subjects which have a wide range of marks have greater weight than those which provide a relatively small scatter around the average mark. Marks can be made to have equal weight by making the scatters the same. The scatter of marks may be expressed by the standard deviation, and a standard score is the distance of a mark from the mean or average of the marks in terms of standard deviation,

$$S = \frac{X - M}{\sigma_x}$$

where X is the raw score, M is the mean mark, σ_x the standard deviation and S the standard score.

(c) *Percentage Mark*

For long, the percentage mark has been used as a means of expressing the results of examinations, with 50 per cent being regarded as a pass mark and anything below it regarded as failing. This of course is quite an arbitrary decision.

The main use of this method of expressing results is to enable different sets of marks to be compared. For example, it is convenient to express as percentages a score of 38 out of 50 in arithmetic and 42 out of 60 in mathematics, in order to compare performances.

(d) *Percentile*

Another method of expressing a score in relation to other marks is to express it as a percentile, that is, the mark below which a specified percentage of marks fall. For example, P75 = 62 means that 75 per cent of the marks fall below 62, or the 75th percentile is 62.

2. *Categoric Marks*

A pupil's total score may be placed in a category. The normal method describing marks by this method is an A, B, C, D, E classification. Categories of this nature have to be defined and a generally adopted technique is to divide the marks into groups of 5, 20, 50, 20 and 5 per cent of the total group: categoric mark A includes the top 5 per cent of the marks. Another less clearly defined definition of the five categories is, well above average, above average, average, below average, well below average.

3. *Rank Order*

Arranging marks in order of merit is another method of relating a pupil's achievement to that of the others who have been tested. Rank 1 is given to the pupil with the highest score and the subsequent ranks are allocated according to the pupil's place in the order of merit. When several pupils have the same score, the ranks are averaged and each pupil given the same rank. For example, if the 4th, 5th, 6th and 7th pupils have the same score each is awarded a rank 5.5. Similarly, identical scores for pupils who are placed 8th, 9th and 10th in order of merit result in each being awarded the rank of 9.

A rank indicates the level of performances of a pupil only in relation to the other pupils sitting the test. If the ability level of the group is low, a rank of 5 may indicate a low level of achievement.

NORMS

When a test is for a specific age group following a clearly defined syllabus, it is possible, by setting the test to a fairly large representative sample of pupils, to establish standards of achievement. The simplest method of providing norms is to construct a table giving the percentile or categoric marks for a given score and age. For example, the achievement of a pupil of the age of 16 years 2 months who scores 32 on a specific standardized test can be interpreted by reading from a table into which percentile or category the pupil falls. Standardized tests of this nature must be very carefully constructed.

Hence there are several methods of expressing the results of a test, but their intended use will determine to a large extent the one chosen.

2. The Purpose of Examinations in Mathematics

As has been stated, examinations are an essential part of all teaching in order to assess the success with which the learning experiences have been organized to attain specific objectives. The assessment of ability and aptitude, achievement, potentiality and effectiveness of teaching is a continuous process bound up with the pupil's progress. The nature of the particular subject will determine what form and with what frequency the examinations and tests will occur.

In mathematics, perhaps more than in other subjects, there is, in most aspects of the work, a continuous development from simple, straightforward ideas to more difficult and complex ones. Without a firm grasp of the former, for example manipulative skills, it is impossible to master later techniques, and further progress is not possible. Hence testing at frequent stages is essential to indicate to the teacher that the pupils are ready to proceed to the next section of the work.

At one time, mathematics examinations followed a traditional pattern. For example, a geometry question would take the form of proving a theorem followed by an example on the theorem. The first part was pure memorization, and the second part had the limited objective of being a direct application of the theorem. Good teaching and reasonable application by the pupil led to apparent achievement in the most unimaginative and the least productive part of mathematics which was the easiest to examine.

RECENT DEVELOPMENTS

The situation now is rather different, due to developments and changes in the content, learning and teaching of mathematics. The movement towards the reconstruction of mathematical education has been world-wide, with many countries employing experimental syllabuses with modern materials and methods. There were several reasons for the starting of this 'new mathematics':

1. The need to integrate the separate compartments of arithmetic, algebra and geometry resulted in the introduction of elements of mathematics not previously included in school courses, for example the language of sets, number systems, vectors and matrices. The importance of *structure* in all mathematics is emphasized by clarifying these and other unifying themes from the beginning of the study of the subject. In this way it is hoped that mathematics will be seen to be more than the teaching and learning of procedures and techniques without the understanding of what is happening and why.

2. The area of applicability of mathematics has extended to such a degree that the purpose of teaching mathematics, defined to be the provision of experiences that help pupils to make progress towards the attainment of certain objectives, has required considerable revision. Knowledge and skills in mathematics occupy a central position in many tasks and activities of present-day society, and it is recognized that the learning of mathematics lies at the basis of improving scientific and technical education. The demand for greater mathematical competence of all citizens also justifies the need to teach mathematics to *all* pupils. The inscription over the door of Plato's Academy forbidding entrance to the man ignorant of mathematics is very relevant to the door of success in fields such as engineering, medicine and genetics.

3. In an attempt to induce a more practical understanding of what mathematics really is, and of the ways mathematicians go about their everyday work, radical changes in teaching methods have taken place. Pupils are given more opportunity to learn by 'discovery', using heuristic and inductive methods as opposed to highly formalized instructional techniques.

At all stages of school this development of new mathematics and new teaching and learning experiences is taking place in such a way that it is difficult to separate their effects. These developments have led to problems in examinations and evaluation, and hence a need to review the different role to be played by them in contemporary mathematics.

Examinations are not a separate entity; they fit naturally into a modern course of mathematics and can be applied to evaluate teaching methods, curricula and learning materials.

If one of the chief roles of present-day mathematics is to maximize the use of mathematical potential, it is essential to determine where the potential is and how it may best be developed. This enquiry must start in the area of formal schooling, and examinations become an indispensable part of the educational process in order to obtain an objective and reliable measurement of mathematical ability.

PURPOSE OF EXAMINATIONS IN MATHEMATICS

In general school testing takes three forms: firstly, the test given at frequent intervals to assess the effectiveness of the teaching of different parts of the work; secondly, the end-of-term or end-of-session test to measure progress made on a complete section of the syllabus; and thirdly, the external examination. Each type can serve to illustrate individual difficulties and points where the selected learning experiences have been successful or otherwise. The nature of and stage reached in a topic will determine the type of questions, for example to test the ability to recall facts or to assess thinking, or to solve problems. Regardless of the type of examination, however, it is vital that the teacher should analyse what the candidate is expected to do to succeed in the test.

1. *The Short Classroom Test*

The prime purposes of this test, which is organized by the teacher himself as part of the educational course, are

(i) to measure the effectiveness of the learning experiences provided for a small part of the course and hence to improve them, and

(ii) to diagnose weaknesses in the pupil's progress.

In the process of discussing the test and correcting errors, areas of misunderstanding and wrong procedures can be located and correct learning reinforced. The test is thus an instrument of teaching, part of the teacher's stock-in-trade, a stimulus for pupils and a means of obtaining a provisional assessment of achievement if required. Such tests are essential and desirable in order to furnish the teacher with information about pupil performance and progress which will help in the process of teaching.

An analysis of a test may suggest changes in objectives, learning experiences, methods or materials. It can help to indicate what content requires to be gone over again, eliminated or revised. Questions designed to test particular mathematical objectives, if correctly analysed, can determine if errors are due to poor learning by the pupils or poor teaching by the teachers. If the learning experiences have, for example, been shown to be inadequate, then they can be reviewed and in consequence help to improve future teaching.

Several types of tests may require to be designed to achieve particular purposes, for example to indicate which pupils at a given time have mastered the concepts and skills of a particular piece of work and are ready to proceed to the next, or to help diagnose the kinds of difficulties pupils are meeting to enable teaching efforts to be better directed.

The test may be asked orally, it may be written on the blackboard or it may be in the form of a written paper. It may take the form of objective questions, conventional questions, short answer questions or a mixture of all types. In general, however, the questions will be short and numerous if testing one particular stage in a piece of work, especially at elementary level, or a smaller number of more complex items, possibly covering one topic in all its aspects if testing a larger or complete section of mathematics. Objective items are useful in order to test one idea at a time, as illustrated in the second of the following examples.

Example 2.1

In testing logarithms the main sources of errors are usually made in

(a) looking up the logarithm and antilogarithm tables
(b) writing down the characteristics
(c) adding, subtracting, multiplying and dividing logarithms, especially when negative characteristics are involved.

Hence to help the teacher to detect where a revision of

teaching experiences is necessary, questions testing each aspect separately, for example:

(i) Give the mantissa (or decimal part) of the logarithm of 6, 285, 4.01, 0.00872, . . .

(ii) Give the characteristic (or integral part) of the logarithm of 600, 23.4, 0.0089, 0.302, . . .

(iii) Give the logarithm of 3600, 3.14, 0.105, 0.000 986, . . .

(iv) Give the number whose logarithm is 1.37, 0.051, $\bar{3}$.078, $\bar{5}$.113, . . .

(v) Using log 5.12 = 0.709, write down the logarithm of 512, 51 200, 51.2, . . .

(vi) Add the following logarithms:

$$3.5 \quad \bar{3}.5$$
$$1.6, \quad 1.7, . . .$$

Subtract the following logarithms:

$$1.2 \quad 1.6 \quad \bar{4}.7$$
$$3.4, \quad \bar{2}.4, \quad \bar{3}.9, . . .$$

Multiply the following logarithms:

$$\bar{1}.5 \times 3, \quad \bar{2}.7 \times 8, . . .$$

Divide the following logarithms:

$$\bar{6}.72 \div 3, \quad \bar{2}.78 \div 3, . . . ,$$

would be more useful than questions such as, evaluate

$$\frac{62.5 \times \sqrt{(0.823)}}{29.2^2 \times 0.000\ 52}$$

which, although picking out pupils who can tackle all the aspects of logarithms correctly, requires close and time-consuming examination to reveal errors.

Example 2.2

This example, in contrast to Example 2.1, illustrates more complex questions which might be given to a class on the completion of quadratic graphs.

The diagram shows the graph of a certain quadratic function f.

(i) The formula which defines the function f is given by $f(x)$ equal to

A $(4 - x)(1 + x)$

B $-4x + 1$

C $x^2 + 3x + 4$
D $(4 + x)(1 - x)$
E $(x + 4)(x - 1)$

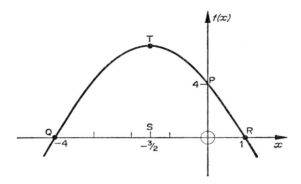

(ii) 1. one zero   2. two zeros   3. a minimum value   4. a maximum value

This function has

A   1 and 3
B   2 and 3
C   2 and 4
D   3 only
E   2 only

(iii) What is the solution set of $f(x) \geqslant 0$?

A $\{x: x \leqslant 4\} \cup \{x: x \geqslant 1\}$
B $\{x: -4 \leqslant x \leqslant 1\}$
C $\{x: x \leqslant 4\} \cap \{x: x \geqslant 1\}$
D $\{x: -4 \leqslant x \leqslant -\frac{3}{2}\}$
E $\{x: -4 < x < 1\}$

(iv) What is the sum of the lengths OP and ST?

A $2\frac{1}{4}$
B $4\frac{3}{4}$
C $7\frac{3}{4}$
D $10\frac{1}{4}$
E   Cannot be determined

For any type of written test it is essential for the teacher to have the main purpose of the test in mind and then design the test to measure the success of the pupils in attaining the objectives for this purpose. For example, are the questions testing for mastery of facts, testing for difficulties in understanding, or required to differentiate between candidates over a large range of ability? Teacher-made tests often emphasize the first of these, assuming that if a pupil knows the facts about a topic then he also understands it. Measurement tends to be confined to objectives that can easily be assessed.

Although pupils may not be able to give reasons for processes in arithmetic and algebra they can often carry out the processes perfectly correctly. Questions such as the following, on the topic of simple equations, can indicate to the teacher, by the responses, whether or not the learning experiences have been successful in achieving the objective of understanding, as opposed to remembering the mechanical process.

Example 2.3

(i) Which of the following are equations?

1. $3x + 4 = 7$
2. $3n + 4 = 4 + 3n$
3. $3 \times 6 + 4 = 22$
4. $3 \times 6 + 4 = 30$

A   1 only
B   1 and 2 only
C   2 and 3 only
D   3 and 4 only
E   Some other response or combination of responses

(ii) An operation ∗ is defined on the positive integers and it is known that {3} is the solution set of $x*x = 20$. Which of the following statements about 4∗4 must be true?

1. $4*4 > 20$ 2. $4*4 = 20$ 3. $4*4 < 20$ 4. $4*4 \neq 20$

A   2 only
B   1 and 4 only
C   3 and 4 only
D   4 only
E   None of them

There are, of course, certain key facts and operations in any mathematics syllabus which must be known and performed automatically and accurately, otherwise further work dependent on this knowledge cannot be attempted successfully. In the case of a test to measure this mastery of factual material, the questions would be mainly of knowledge and comprehension and the scores should be high, confirming that the teaching and learning has been successful. If, however, the test is to assess the ability to solve problems or to assess the understanding of concepts and principles, then the questions would have a different weighting, the emphasis would be on items involving unfamiliar situations and in consequence more errors would be expected, also providing useful diagnostic information. It is much more difficult to test these higher abilities.

Example 2.4
The following set of examples illustrates the levels of learning in the case of the associative law.

(i) Which of the following is an example of the associative property of multiplication?

1. $6 \times (7 \times 8) = 6 \times (8 \times 7)$
2. $6 \times (7 \times 8) = (6 \times 7) \times 8$
3. $6 \times (7 + 8) = (6 \times 7) + (6 \times 8)$

A 1, 2 and 3
B 1 only
C 2 only
D 3 only
E Some other combination of 1, 2 and 3

(i.e. pupil must be able to recognize a special case of the law).

(ii) State, in symbols, the associative property of addition (i.e. pupil must be able to state the law in symbols).

(iii) For which of the following definitions of * on the set of whole numbers is * associative?

1. $x*y = 2x + 2y$
2. $x*y = x - y$
3. $x*y = 2xy$

A   1, 2 and 3
B   1 only
C   2 only
D   3 only
E   Some other combination of 1, 2 and 3
(i.e. pupil must be able to apply the law to a new problem).

(iv) Invent some operations which are associative and some which are not (i.e. pupil must be able to use law to develop a new concept).

It is necessary for teachers to be able to identify and evaluate such levels of understanding as are considered appropriate.

To set such questions the teacher must have the mathematical objectives of the syllabus clearly defined. Instructional objectives are necessary in order to devise learning experiences to attain them, and the purpose of the test is then either to indicate to the teacher that the experiences have been successful or to point out at which stage they have failed. For example, if a syllabus states that 'knowledge of surds' is a topic then the teacher has no basis for providing learning experiences. What determines when a pupil 'knows'? What must a pupil be able to *do* in order to show that he knows? In such a case the teacher must rewrite the statement in order to provide a foundation for the learning experiences and the test to be constructed to check on their effectiveness. The objectives might then include 'to be able to simplify surds using the property $\sqrt{(ab)} = \sqrt{a} \times \sqrt{b}$,' and 'to be able to add, subtract and multiply surds'. If no further objectives are specified then the teacher can design questions to tell whether the pupils have reached a particular objective and also will avoid wasting time devising unnecessary learning experiences and tests, for example for the ability 'to be able to divide surds'.

In general, marks in a classroom test are primarily intended to give information about each pupil's progress and not to produce a rank order, but after a series of tests they could be used to grade a large number of pupils into broad groups especially in the first two years of the secondary school. Obvious essentials to this testing are quick administration and rapid marking, and objective questions fit these requirements ideally. Objective items have also been proved to be a powerful aid to diagnosis, one of the prime functions of a school test, and it is important that teachers follow the development of this type of testing in the field of examinations.

If the correct answer to a certain objective item is 'E' but the most popular one is 'C', then the teacher must consider why so many pupils made that specific error. The answer may be straight-forward (not enough time spent on the work), more complicated (learning experiences correctly selected but presented in the wrong order), or confusion with regard to certain concepts and skills. The following are examples to illustrate this point.

Example 2.5

(i) If $f(x) = \log_b x$ then $f(kx)$ is equal to

A $f(x) - f(k)$
B $kf(x)$
C $f(k).f(x)$
D $f(x)$
E $f(k) + f(x)$

(ii) If $9^9 = 3^x$ then x is equal to

A   9
B 3^9
C   27
D 2^9
E   18

(iii) If the functions f and g are defined by the formulae $f(x) = 3x + 2$ and $g(x) = x - 3$ then $f \circ g$ is defined by

A $3x^2 - 7x - 6$
B $4x - 1$
C $3x - 1$
D $2x + 5$
E $3x - 7$

In each case E is the key. Suppose that C is the most popular answer. There are several possible deficiencies in understanding suggested by the incorrect answers which would be of use to teachers in assisting pupils in learning mathematics more effectively.

Another important purpose of the classroom test, especially with non-certificate pupils, is to verify to the teacher that his pupils are numerate, as numeracy and literacy are two essentials in all education. Diagnostic tests, in particular, would be useful in this context by providing information on pupils' strengths and weaknesses which would otherwise be difficult to pinpoint.

School tests of the type just described, and the uses made of their results, are often criticized, but they can be made interesting, informative and diagnostic, and the advantage they have is that they can be used for instructional as well as evaluation purposes. One criticism of testing, for example, is that there is a loss in teaching time. This is more than balanced, especially in mathematics, by the fact that the testing results in a strengthening and revision of the main points of difficulty and of unsuccessful learning experiences which have been pinpointed by the test.

In attempting to measure more than basic computational skills, as the teacher must at some stage, much time and energy will be necessary, but test construction can be a learning experience for the teacher as well as a method of re-evaluating objectives and effectiveness of teaching. There are three points which must be emphasized with regard to classroom tests:

(a) they must be designed for specific purposes,
(b) items and questions must be constructed which measure objectives other than remembering facts,
(c) careful analysis of results, as well as improving learning by identifying errors, can improve the tests by enabling a pool of successful questions to be built up, each of which measures what it was intended to measure.

2. *The Terminal Test*

This test can be, and normally is, used to serve many purposes. The prime purpose is to assess achievement in mathematics over a term or a session and consequently it may be longer and more complex than the informal classroom test.

In order to assess achievement in mathematics it is necessary to define what the pupil has to be able to do at the end of the course. This is extremely difficult to do, but it would appear that there are factors involved other than 'an ability to do mathematics'. There are many pupils, for example, who have an outstanding ability in, say arithmetic or geometry, but only a moderate ability in other branches. Evidence has been produced to confirm that mathematical ability is complex and that the analysis of the processes involved in mathematical thinking is not straightforward. One factor, however, which appears to be necessary is an intellective one. The International Study of Achievement in Mathematics supported strongly the association between intelligence and the score made in the achievement test. The same study compared the performance of boys and girls and found that in almost all the groups tested boys were superior to girls, but that the superiority was much less marked in coeducational schools than in segregated schools, which perhaps indicates that the narrowing of the gap was due to the attitudes adopted towards the ability to learn mathematics. As well as mathematical ability there must therefore be an ability to learn mathematics, since it is unlikely that boys are different in ability from girls.

It would appear, therefore, that achievement in mathematics can be improved by interest in and possibly attitude towards the subject, and another factor is good teaching. Further, most teachers will agree that at a certain level, interest, perseverance and the will to learn are just not enough. Something extra is required to improve performance, a high degree of the innate ability which cannot be taught and is almost impossible to define. It might be called a 'flair' for mathematics, and in no field is the difference between the best and the very good pupils as great. There is that special gift for mathematics which, although it can be cultivated, can only be provided by nature. This ability requires not only the ability to memorize facts and to perform computational techniques but also to make judgments, to think clearly and to discover relationships among previously unrelated concepts as opposed to being able to do routine manipulation of previously learned material, in other words, the difference between reproductive and productive thinking.

If it is difficult to describe mathematical ability, then there must be even greater difficulties in testing it. It follows that the first essential in constructing such a test is a clear definition of the objectives of instruction which are considered to reveal this ability. Achievement will then be measured in terms of *all* the objectives and it is vital that a candidate must have achieved some success in all of them in order to pass this type of examination. This necessitates questions covering the whole syllabus, and as one of the main characteristics of objective items is wide syllabus coverage the means is available.

A second purpose of terminal examinations is motivation. A test is a powerful motive and can exert a considerable influence upon the quality and direction of a pupil's work. What the pupil studies in mathematics to a large extent depends upon what he expects to be tested on, and the manner in which he works is determined by the type of test which he anticipates at the end of the course. The implication is that by improving objectives, types of tests and methods of constructing them, an improvement in learning can result. It might in fact be profitable for teachers to identify the factors of motivation. What motivates one pupil, for example the need to be persuaded that it is necessary to learn mathematics, may not motivate another. Tests can evaluate and identify these factors of motivation. Reinforcement of learning usually occurs when pupils are preparing for a test, resulting in a better understanding of the elements of the course and their interrelation, indicating again the role that tests can play in controlling and helping the learning process.

At practically every stage at which a decision has to be made about a pupil's mathematical future, selection is involved. Terminal examinations can serve this purpose, supplemented by classroom tests and other relevant assessment procedures. The decision is usually based on the answers to two questions about the pupil:

(a) What is his attainment in mathematics to date?
(b) What is his potential to do mathematics in the future?

In other words, evidence of what has been achieved and what is likely to be achieved is sought. Hence, certain tests must be constructed with the purpose of prognosis in mind if they are to be of use in helping to make a decision in which question (b) is involved. In such tests questions which differentiate effectively between pupils are necessary, and in the first three years of secondary school the results can be used to set and/or stream pupils in mathematics, to guide pupils in their decision to continue the subject or otherwise in a subsequent year or, to help future employers if no external examinations are available, to compare applicants from the same school. If the prognostic purpose of a test is success in a later school course in mathematics then the extent to which it achieves its aim can be measured over several years, giving a check as to whether the test is useful for this purpose or not.

3. *External Examinations*

In a similar way to school terminal examinations, external examinations in mathematics can serve several purposes. For example, they can serve to assess what has been achieved and what is likely to be achieved in mathematics. Inevitably all of these examinations are used for both purposes, but if the main aim is qualification and the extent to which a pupil has benefited from the mathematical education he has received, the questions will differ from those in an examination constructed to select for or predict success in, for example, a mathematics course at university or some role in society. Consequently, to suit both purposes questions are included which (a) measure what should have been achieved at the end of the course and (b) differentiate between pupils and measure aptitude. Recent developments in education tend to put the emphasis in examinations on (a), whereas the university and future employers are concerned with placing the emphasis on (b).

External examinations provide motivation to both pupils and teacher in the form of learning and teaching respectively. They offer a possible criterion by which a teacher can judge the progress of his own pupils, and over several years can serve to indicate changes in his effectiveness as a teacher reflected in the results of his pupils, and to compare his standards with those of other teachers. External examinations have their critics, but without some common examination the selection of candidates to fill a limited number of university places cannot be satisfactorily determined. Research in America has shown that the Scholastic Aptitude Test, an objective-type test, part of which is a mathematics paper, is a good predictor of success in following a university course.

CONCLUSION

Examinations in mathematics are not an end in themselves. The end is to increase the effectiveness of the educational process. To achieve this, tests must be devised, the results of which can be related to and be used to improve this process without impeding the pupil's general mathematical development. Objective testing is an essential component of such tests and examinations, and if properly constructed has a less harmful influence on the educational system than traditional types of testing.

Examinations in mathematics may be designed to fulfil one or more of a wide variety of purposes, but it is essential to be as specific as possible about the main functions of a particular examination in order to avoid the danger of assuming that one type of examination will effectively meet several purposes.

3. *Educational Objectives*

INTRODUCTION

Confusion may arise when objective testing and educational objectives are being discussed, since the word 'objective' has a different meaning in the two cases. The objectivity in the former arises from the objectivity of the marking process, compared to the less objective scoring of most other assessment procedures. It is not possible, however, to discuss objective testing without reference to the educational objectives of the course. In this sense an objective is the development of a behaviour pattern which teachers and examiners are prepared to accept as evidence of learning, or is comprised of a set of explicit statements of the ways in which pupil behaviour is expected to be changed by the learning process, or is made up of a collection of observable 'terminal' behaviours. In attempting to describe these 'desired changes in behaviour', a value judgment is required to decide what changes are desirable and the behaviours require to be observable and in some degree measurable.

AIMS AND OBJECTIVES

Syllabuses tend to be vague and often give the impression that secondary education is primarily concerned with covering topics. The topics are usually given in broad terms with respect to specific content areas, with little indication of the depth of treatment desired, and consequently many interpretations are made by teachers. For example, most mathematics syllabuses contain statements such as 'the properties of common shapes', 'the sphere', and 'proportion and scale'. Most syllabuses also give the broad aims of a course which are essential to give emphasis and purpose to it, indicating the outcomes of learning in a particular subject considered desirable from personal and social considerations. What must also be given, however, after consideration of the aims, are many statements in behavioural terms indicating the skills and abilities to be acquired or developed by the pupils in attaining these aims, and hence to be tested by the examination and other evaluation procedures. These are the objectives of the course, and as is obvious from this an aim and an objective are being given different meanings. For example, an aim of a mathematics course might be to emphasize the cultural value of mathematics or 'to interest the pupils in mathematics', which on paper appears impressive but in fact provides little direction with respect to devising learning experiences for the pupils. Whereas if an objective is knowledge of some kind then precise descriptions, such as 'the pupil, at the end of the course, should be able to state the definition of an acute angle, or be able to write the commutative, associative and distributive laws for addition and multiplication of real numbers', will be available to the teacher in order to determine and select appropriate experiences to attain the objective.

CHANGES IN BEHAVIOUR

It is essential to consider the changes that are expected in the pupils at the end of a course in terms of remembering, comprehending, applying etc., with respect to certain content areas. These changes in behaviour are often broken down into three domains:

(a) changes in the 'thinking' area—called the cognitive domain;
(b) changes in the 'feeling' area—called the affective domain;
(c) changes in the 'acting' area—called the psychomotor domain.

Changes in (a) will result in the acquisition of knowledge and the development of those skills and abilities necessary to use knowledge, e.g. the ability to solve problems in mathematics. Changes in (b) will be recognized by, for example, an interest in or an appreciation of the subject at the end of the course which was not initially there. Changes in (c) will result from development in manual and motor skills, and much has still to be discovered about this domain. An example in mathematics would be the development of skill in using instruments or in sketching geometric figures.

If the educational objectives of a course are laid down in the form of human behaviours then it is much easier to measure the success of pupils in reaching them; also, if each ability is in the form of an objective, then the task of constructing learning experiences to achieve the objectives, which must always remain in the hands of the teacher, will also be clearer.

CLASSIFICATION OF OBJECTIVES

Aims of a course are usually few in number but objectives are many and due to the importance attached to their definition and specification, experts in the field of measurement have found it useful to classify and identify the latter statements under main categories of behaviour. The standard work on classification of educational objectives, described in operational terms, is the *Taxonomy of Educational Objectives*, in particular Handbook I (Bloom, B. S., *et al.*, 1956), which deals with categories of objectives in the cognitive domain. It is designed to be a classification of the pupil behaviours which represent the intended outcomes of the educational process. Each category is analysed together with illustrative objectives and test items.

For the purposes of classification the cognitive objectives are divided into two parts:

1. objectives which involve the acquisition and recall of knowledge;
2. objectives which involve the development of those intellectual abilities and skills necessary to use knowledge.

The following is a summary of the objectives, placed into categories as indicated by the Taxonomy.

1. *Knowledge*

1.00   *Knowledge* (Remembering previously learned material)
  1.10   Knowledge of Specifics
    1.11   Knowledge of Terminology
    1.12   Knowledge of Specific Facts
  1.20   Knowledge of Ways and Means of dealing with Specifics
    1.21   Knowledge of Conventions
    1.22   Knowledge of Trends and Sequences
    1.23   Knowledge of Classifications and Categories
    1.24   Knowledge of Criteria
    1.25   Knowledge of Methodology

    1.30   Knowledge of the Universals and Abstractions in a Field
      1.31   Knowledge of Principles and Generalizations
      1.32   Knowledge of Theories and Structures

2. *Intellectual Abilities and Skills*

2.00   *Comprehension* (Grasping the meaning of material without necessarily relating it to other material)
  2.10   Translation
  2.20   Interpretation
  2.30   Extrapolation
3.00   *Application* (Using the information in unfamiliar concrete situations)
4.00   *Analysis* (Breaking down material into its constituent parts)
  4.10   Analysis of Elements
  4.20   Analysis of Relationships
  4.30   Analysis of Organizational Principles
5.00   *Synthesis* (Putting parts together to form a structured whole)
  5.10   Production of a Unique Communication
  5.20   Production of a Plan, or Proposed Set of Operations
  5.30   Derivation of a Set of Abstract Relations
6.00   *Evaluation* (Judging the value of material for a given purpose using definite criteria)
  6.10   Judgments in Terms of Internal Evidence
  6.20   Judgments in Terms of External Evidence

The taxonomy represents a hierarchical order of the different classes of objectives with the subdivisions also being in order of increasing complexity. The objectives in one class are likely to make use of and build on the behaviours found in the preceding classes in the list. The educational behaviours are arranged from simple to complex, the idea being that a particular simple behaviour may integrate with other equally simple behaviours to form a more complex behaviour. This can be represented as shown below.

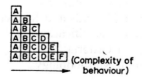

(Complexity of behaviour)

Each category is more complex than the previous one and builds upon it although there may be some overlap between categories. Hence Knowledge is the lowest ability, followed by Comprehension and Application, which are higher abilities than knowledge since they are dependent on it. When testing objectives which fall into the categories of Analysis, Synthesis and Evaluation, the 'higher abilities', mastery of related material in the lower categories is assumed.

The taxonomy is intended to have universal application and hence it has to be adapted accordingly when the subject specialist attempts to analyse the objectives of a course, for example, what is meant by the category of Application with reference to Mathematics? The taxonomy also does not indicate the relative importance of the objectives. This will depend upon several factors, for example the subject and level of the pupil aimed at. Despite the problems of interpretation, a thorough acquaintance with the classification is invaluable in item writing and examination construction.

There are other simpler and cruder classifications by type of objective, for example the following categories were found to be more relevant to particular course specifications and needs at the College of Medicine at the University of Illinois (1962):

1.0 Knowledge
2.0 Generalization (Ability to select a relevant generalization to explain specific phenomena)
3.0 Problem-solving of a Familiar Type
4.0 Problem-solving of an Unfamiliar Type
5.0 Evaluation (Ability to evaluate a total situation)
6.0 Synthesis

EDUCATIONAL OBJECTIVES AS PART OF CURRICULUM PLAN

Once the educational objectives of a course have been fixed there are other vital components of the complete process to be considered:

(a) the construction and selection of learning experiences to achieve the objectives;
(b) the measurement of the extent to which the objectives have been achieved, i.e. a means of evaluating the success of

pupils in attaining the objectives laid down at the beginning of the course;

(c) on the basis of (b), a modification of the choice of the objectives and the learning experiences.

This can be shown diagrammatically with the four components interacting to give a curriculum plan. Each component has its place in the plan but must be thought of in terms of the others

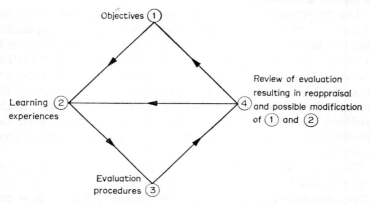

and not on its own since each has a possible beneficial influence on the others. In this way evaluation, which should be a continuous process and not just an examination at the end of the term or course, can be constructed to test the chosen objectives and learning experiences, and their success in bringing about the required behaviour changes in the pupils. Similarly the objectives can develop both learning experiences and evaluation techniques, and the learning experiences can clarify objectives and suggest methods of evaluation.

At present, especially in Certificate courses, the examination dominates both the teaching methods and the learning experiences, due to the fact that the objectives of the courses are not clearly stated in terms of the kinds of behaviour the pupil is expected to exhibit if he has achieved them, and consequently the teacher and the examiner may interpret the meaning of some part of the syllabus in different ways. The examination is left in isolation, either testing the wrong objectives or bad teaching due to poorly constructed learning experiences. The ability that the teacher develops in the child in a particular topic, for example the

Laws of Indices, may not be the ability that the examiner tests in the examination. To be useful, tests or examinations must measure performance in terms of the objectives of the syllabus.

R. F. Mager, in his book *Preparing Instructional Objectives*, indicates rules to be followed when writing objectives which will describe the desired behaviour of the learner, for example, 'Write a statement describing the objective and then modify it until it answers the question "What is the learner doing (or doing better) when he is demonstrating that he has achieved the objective?" ' To develop an appreciation for music is a vague objective, whereas the ability to be able to solve quadratic equations by factors, completing the square or by the formula is a well-defined objective. To understand the laws of set algebra is a vague objective since it does not indicate what the pupil will be doing when he understands the laws of sets. Does he have to be able to say them, or list them or use them in solving set problems etc? Words to be avoided in stating objectives are words like 'understand', 'know' and 'appreciate'. Words to be used instead are 'identify', 'solve', 'list' and 'contrast', since they are more meaningful and specific, as well as indicating observable behaviours.

It is also important to describe the conditions under which pupils' behaviour is expected to occur. This involves explaining what information will be given and can be illustrated via a sample question. For example, the pupils must be able to factorize the following types of expressions: $ax + bx$, $a^2 - b^2$, $ax^2 + bx + c$. Statements of objectives, if they are to be of any value, must spell out in detail observable changes in a pupil's behaviour at the completion of the course. In this way the specific objectives become readily examinable with objective-type items especially suitable for this task.

Generalized statements of objectives such as 'acquiring an appreciation of mathematics' receive little attention in the classroom because teachers cannot devise learning experiences to enable pupils to attain such objectives. In attempting to find behaviours associated with 'appreciation', the following are found: 'wants to know more about', and 'attempts to create more examples of', with the appropriate content to complete the statements. Furst suggests that

(a) objectives should be stated clearly in terms of behaviour and at the correct level of generality;

(b) care should be taken to avoid overlap when listing them;
(c) they should specify the kinds of responses that may be accepted as evidence of the aspects of behaviour related to the problem of evaluation;
(d) they should specify the limiting conditions under which these responses are likely to take place.

It is essential that examinations test abilities of educational value and not just a series of topics written into a syllabus. Without clearly defined objectives specifically written in terms of desired behavioural patterns, this can very easily happen.

The starting point is to lay down the instructional objectives. This presumably must be done by the subject panels responsible for the formation of the curriculum, or if teachers produce their own syllabuses and examinations then the onus will pass to them. Subject specialists must define abilities which they think are able to be developed by their subjects and to be educationally worthwhile teaching objectives. These will vary depending upon the age and experience of the pupils and may be weighted according to what the terminal or external examination is designed to test, for example the measurement of attainment, prognosis, diagnosis or selection. These objectives would then be available for teachers, who would have to construct suitable learning experiences and teaching methods in an attempt to reach the goals. The evaluation process would then test the success of both the set of objectives and the learning experiences.

SPECIFICATION OF AN EXAMINATION

The written examination which may be part of the evaluation process requires at least two aspects of specification, (a) the abilities to be tested and (b) the subject matter. A two-dimensional grid, usually called the blueprint, can be drawn up in which the objectives of the course are plotted against the appropriate content areas.

The grid shown is for a 100-item examination in which the totals down the right-hand side indicate that content areas 1 and 2 are to receive more emphasis than areas 3 and 4. The totals along the bottom indicate, for example, that 15 per cent of the items are designed to test ability D.

This type of blueprint will help to ensure that the correct

Objectives Content area	A	B	C	D	E	Totals
I	12	5	8	5	—	30
2	8	5	11	4	2	30
3	5	4	7	2	2	20
4	5	6	4	4	1	20
TOTALS	30	20	30	15	5	100

weighting is given to the various abilities and content areas. This will normally depend upon the purpose of the test, the age group being taught and the time spent on particular sections of the syllabus during the school session, but in addition it is essential that it reflect the emphasis given and importance attached to topics while they are being taught. Such an exercise may help teachers when they are constructing an examination, to guard against a tendency to include only those types of teaching experiences in their course that can be taught and tested most easily, for example to attain knowledge of specific facts. With the introduction of new syllabuses in many subjects teachers now have the opportunity to avoid emphasizing the lower ability objectives which were prevalent in traditional examinations. It is also true to say that many pupils limit their studying to the types of learning which they believe the teacher will include in the next examination, and hence an over-emphasis on knowledge objectives can have a harmful effect on pupil achievement. In mathematics, for example, the acquisition of basic facts is important, but a course in mathematics should result in much more, with the development of logical reasoning, creativity and positive attitudes towards the subject, requiring greater emphasis.

An added effect of the blueprint is that teachers are forced to study each topic in the syllabus and consider whether it is of use in developing a required objective. This will enable unnecessary material to be weeded out from the syllabus, thus providing a valuable feedback effect. The above type of grid can be used for both essay-type and objective-type examination papers but lends itself more easily to the latter, since the large number of items that can be used in this type of testing allows questions to be devised which specifically attempt to measure one particular objective and this makes use of one of the advantages of the blueprint, namely that each question of the examination is uniquely defined with respect to behaviour and content area.

OBJECTIVES IN OTHER DOMAINS

In some subjects, for example mathematics, most behaviours seem to have cognitive origins, and with an adaptation of Bloom's Taxonomy and a blueprint of the above type a reasonable evaluation procedure can be visualized. An evaluation process for assessing changes of behaviour in the other domains is perhaps more difficult to produce. For example, how is evidence of a change of attitude towards or a development of an appreciation for a subject obtained? Another example is that of the problem facing primary school teachers who must also list objectives of their teaching. Such a list might include, (a) knowledge, (b) comprehension, (c) values, for example social and moral, and (d) originality, among others. How does the teacher assess social and moral values? A possible answer is by a series of observations made over a period of time, which can be recorded with the results of performance made using other evaluation procedures to assess (a), (b) and (c), in some form of assessment record card. The point is that if certain objectives are deemed relevant in a course then they require to be evaluated regardless of the domain into which they fall. An objective examination may not be the ideal answer to evaluate a certain objective, and in such a case a practical, oral or some other assessment procedure will be required. A combination of several techniques may be the answer, for example the following has been used by several colleges of medicine in America as a means of obtaining a 'performance profile' of each student:

   (a) a written comprehensive examination (almost totally objective)
   (b) a practical examination in laboratory and clinical skills, supplemented by

(c) descriptive ratings provided by individual instructors that assess motivation, attitudes, work habits and other elements of behaviour that enter into a student's performance.

Phillips lists some of the tools and techniques used in measuring and evaluating objectives not restricted to the cognitive domain. Examples are,

(a) a sociogram—a chart which portrays the social relationship of individuals in a particular group,
(b) a questionnaire—to measure attitudes and opinions.

In Scotland, Curriculum Paper No. 7, *Science for General Education*, H.M.S.O., attempts to specify objectives in the attitudinal field, acknowledging that learning experiences may be successful in attaining objectives in the cognitive domain but fail to increase a pupil's interest in or appreciation of the particular subject being taught.

The importance of the affective domain has been recognized, however, and a taxonomy of such interest and attitude objectives has been produced, the point being that most cognitive objectives have an affective component. While the pupil is acquiring knowledge and skills there are also taking place concomitant learnings in attitudes, appreciations and interests. The categories of objectives in the affective domain are as follows, again arranged in hierarchical order.

1.0 *Receiving* (attending)
   1.1 Awareness
   1.2 Willingness to Receive
   1.3 Controlled or Selected Attention
2.0 *Responding* (describes 'interest' objectives)
   2.1 Acquiescence in Responding
   2.2 Willingness to Respond
   2.3 Satisfaction in Response
3.0 *Valuing*
   3.1 Acceptance of a Value
   3.2 Preference for a Value
   3.3 Commitment
4.0 *Organization*
   4.1 Conceptualization of a Value
   4.2 Organization of a Value System

5.0 *Characterization by a Value or a Value Complex*
  5.1 Generalized Set
  5.2 Characterization

The problem in measuring change in the affective domain is that some of the objectives of teaching are not amenable to testing in a written examination. However in learning mathematics, for example, pupils will develop attitudes towards some aspect of the subject and assessment of these attitudes can serve a useful purpose in the educational process.

There have been other attempts to classify objectives in the affective domain, for example Gronlund suggests the following types of cognitive and non-cognitive outcomes: (a) Knowledge, (b) Understandings, (c) Thinking Skills, (d) General Skills, (e) Attitudes, (f) Interests, (g) Appreciations, (h) Adjustments.

Little work on objectives in the psychomotor domains has been done although it is currently being developed. The area contains objectives such as manipulation of material and objects, and motor skills are important for certain vocational activities, in particular for those that depend on speed of response, for example typists and machine operators. Subjects such as science, geography and technical studies must be taught with objectives related to the development of practical skills and abilities in mind. A taxonomy of objectives in the psychomotor domain would be useful to suggest appropriate learning experiences to achieve such skills. R. H. Dave has proposed the following hierarchical classification of psychomotor behaviour in the hope of initiating thinking in this direction. The common factor in the following objectives is based on the concept of co-ordination.

1.0 *Imitation*
  1.1 Impulsion
  1.2 Overt Repetition
2.0 *Manipulation*
  1.2 Following Direct
  2.2 Selection
  2.3 Fixation
3.0 *Precision*
  3.1 Reproduction
  3.2 Control

4.0 *Articulation*
  4.1 Sequence
  4.2 Harmony
5.0 *Naturalization*
  5.1 Automatism
  5.2 Interiorization

CONCLUSION

It is essential to clarify, at the beginning, the educational objectives of a course, and for the teacher to be familiar with them before devising and arranging learning experiences for the pupils. If examination boards responsible for external examinations can provide lists of aims and detailed information on the topics of a course then the logical next step, for the boards or the teachers if necessary, is to consider the second dimension and produce statements of behaviour to indicate the precise objectives necessary as a basis for the production of learning experiences to help pupils to achieve these objectives. The syllabuses must change from being 'examining' syllabuses to become 'teaching' syllabuses.

Objective testing can be a valuable asset to teachers in constructing examinations to match the objectives, both present and future, and consequently help to improve the educational process as a whole.

4. *Educational Objectives in Mathematics*

The previous chapter considered educational objectives in general. This chapter deals with objectives in the teaching of mathematics.

A differentiation was made between an aim and an objective of a course and it is relevant to pursue this distinction further with respect to mathematics.

EDUCATIONAL OBJECTIVES FOR A MATHEMATICS COURSE

The following diagram illustrates a classification of the objectives of a course of study.

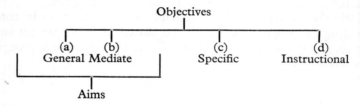

The interpretation of this is that educational objectives in mathematics can be classified into four types or levels, namely: (a) general, (b) mediate, (c) specific, (d) instructional; (a) and (b) are taken by most authorities when discussing objectives, to be the aims of a course. All levels are necessary in the educational process. The first level is required for the development of the second, and similar guidance is necessary in order to list objectives at the other levels.

(a) *General Objectives*

Such objectives are the same for all subjects, and are best described as statements of intent. A list of general objectives will include

(i) the development of the ability to think,

(ii) the development of rational powers,

(iii) the full and harmonious development of the individual,

(iv) to develop a healthy citizen useful for home, society and the nation.

These are necessary since the subject of mathematics cannot stand outside the general system of education, and must play its part in the attainment of the objectives of education. It is generally accepted that there are three basic general objectives required for any subject wishing a place in the curriculum. These are utilitarian or practical, disciplinary and cultural. With regard to mathematics the first concerns the usefulness of the subject in society, the second concerns the acquisition of certain intellectual skills such as the ability to recognize logical relations and to generalize, and the third concerns, for example, the acquisition of an appreciation of the powers of mathematics. The problem is that the objectives will require re-evaluation and restatement not only from era to era but also from country to country, since they must be dependent upon the predominant economic and scientific needs of a country as well as its distinctive culture and the changing demands of its society.

(b) *Mediate Objectives*

Such objectives are better defined as the broad aims of a course in mathematics. They are common to all mathematics teaching but are vague and completely inadequate for the teacher. All-embracing statements of aims, such as numeracy, a tool for other sciences, or a training in logical thinking, are not sufficient to help in the actual teaching situation but are necessary to justify the inclusion of the subject in the curriculum. Most courses and examination syllabuses in mathematics mention objectives of this type which indicate *why* the subject is being taught to pupils but do not describe *what* the pupils should be able to achieve after the learning process has been taught. An example to illustrate this appears in the pamphlet issued with the books written by the Scottish Mathematics Group. The statement reads, 'The course aims to interest pupils in mathematics and to emphasize its relevance in the world of today.' Another example is taken from the general statement of purpose for a particular series of mathematics tests: 'Their purpose is to measure the growth in those mathematical concepts, abilities and skills considered essential for the mathematical literacy of the average well-informed person, irrespective of his major field of interest.'

Other examples of broad aims relevant to secondary education are:

   (i) mathematics as a means of communicating quantifiable ideas and information;

 (ii) mathematics as a training for discipline of thought and logical reasoning;

(iii) to begin to understand the powers and limitations of mathematics;

(iv) the inculcation of a feeling, almost a love, for mathematics.

In the 1955 Mathematics Association Primary Report, it is stated that the aim of primary mathematics teaching is the laying of a 'foundation of mathematical thinking about the numerical and spatial aspects which children of that age encounter'.

Broad aims will also vary from one mathematics course to another. For example, the aims of a course for pupils in the first two years of secondary school will differ in emphasis from those of a course in mathematics designed for sixth year pupils.

(c) Specific (or Process) Objectives

These are the aims of a course written in behavioural terms in order to make them a worthwhile tool in teaching. They describe the abilities which it is hoped that the course will develop. The general statements of outcome in (a) and (b), which are too abstract and too ambiguous, are replaced by detailed and concrete specifications in terms of observable pupil behaviour which indicates achievement of the objectives.

Process objectives are independent of subject-matter content and it is the task of the curriculum developer and test maker to interpret the categories of Bloom, for example, for his own subject.

(d) Instructional Objectives

These are composed of two parts—*the behaviour* the pupil is to show and the *subject-matter content*, which depends upon the needs and level of development of the pupils, that is used in displaying this behaviour. Bloom's Taxonomy deals only with the first part, giving a detailed and invaluable analysis of the behaviour dimension, but with the categories to be applied in different ways to different subject fields. Hence it is necessary to indicate appropriate content areas through which the behaviour is to be developed and the specific objectives attained. Content material available in examination board syllabuses and textbooks is usually analysed horizontally into areas such as arithmetic, algebra, geometry, trigonometry, etc. For example, an attempt to provide instructional objectives was made by the School Mathematics Study Group in the National Longitudinal Study of Mathematical Abilities. An adaptation of Bloom was made to establish seven levels of intellectual activity in mathematics and a list of units of subject matter which should be tested was selected. An outline of the two classifications is as follows:

Specific Objectives	*Content units*
1. Knowing	1. Systems of Numbers
2. Translating	2. Measurement
3. Manipulating	3. Geometry
4. Choosing	4. Co-ordinate Systems and Graphs
5. Analysing	5. Algebraic Sentences and their Solutions
6. Synthesizing	6. Relations and Functions
7. Evaluating	7. Algebraic Expressions
	8. Probability and Statistics
	9. Logic
	10. Applications

Taking the major content classification as rows of a matrix and the major levels of cognitive behaviours as columns, then the cells obtained give approximate definitions of instructional objectives.

A working set of such objectives is essential for both test constructor and curriculum maker. In the derivation of the objectives by this method the behaviour or process part is taken to be more important than the subject-matter dimension. The first step is to list the behaviours that are judged to be desirable for the course or examination, and next appropriate topics which can be used to bring about these objectives are selected.

Example 4.1—Examples of instructional objectives
1. *Specific Objective:* Comprehension ⎱
   Topic: The system of whole numbers ⎰

Instructional Objective: The pupil should be able to simplify, with or without brackets, expressions involving the four operations.

Item When $10 + 4 \times 3 - 10 \div 2$ is simplified the answer is

  A -7
  B 2
  C 16
  D 17
  E 37

2. *Specific Objective:* Synthesis⎫
  Topic: Number patterns ⎬
Instructional Objective: The pupil should be able to abstract, symbolize and prove the general relationship implicit in particular numerical cases.

Item Consider the following number sentences

  (i) $7(2 \times 2)$ $= 7 \times 2^2$
  (ii) $7(5 \times 5)$ $= 7 \times 5^2$
(iii) $7(\frac{1}{4} \times \frac{1}{4})$ $= 7 \times (\frac{1}{4})^2$
(iv) $7(6\frac{3}{4} \times 6\frac{3}{4}) = 7 \times (\frac{27}{4})^2$

Which of the following algebraic statements is the generalized form of the pattern?

  A $7(3\frac{1}{2} \times 3\frac{1}{2}) = 7(\frac{7}{2})^2$
  B $n(2 \times 2)$ $= n(2)^2$, n a whole number
  C $n(a \times a)$ $= na^2$, n and a whole numbers
  D $n(7 \times 7)$ $= n.7^2$, n a whole number
  E $7(n \times n)$ $= 7n^2$, n a rational number

3. *Specific Objective:* Knowledge   ⎫
  Topic: The system of whole numbers⎬
Instructional Objective: The pupil should be able to recall the properties of the four operations on whole numbers.

Item Which of the following operations with whole numbers will always give a whole number?

I Addition   II Multiplication   III Division

  A I only
  B II only
  C III only

D   I and II only
E   II and III only

4. *Specific Objective:* Technique ⎫
   Topic: Systems of equations of the first degree in two ⎬
   variables ⎭
   Instructional Objective: The pupil should be able to solve systems of equations of the first degree in two variables with numerical coefficients.
   Item   If $x + y = 6$ and $x - y = 2$, then x is equal to

A   0
B   2
C   4
D   8
E   10

As mentioned previously, instructional objectives will depend on the aims of a course and also according to the phase of the school course. Levels of development in the pupils must be recognized as affecting objectives, since a particular objective is only realistic if it *can* be attained.

TAXONOMY OF EDUCATIONAL OBJECTIVES IN MATHEMATICS (COGNITIVE DOMAIN)

Need for a Taxonomy

Until the publication of Bloom's Taxonomy in 1956 there had been no hierarchical classification of educational objectives. Previous attempts had been made to state objectives, but although usually given in the form of levels of performance they were not comprehensively listed from the lowest to the highest. Without such a taxonomy as a basis to formulate objectives the tendency has been, and still is to a great extent, to emphasize only knowledge and comprehension and to avoid the higher mental processes involving understanding and critical thinking. Especially in mathematics, a vital ability to be developed is that of applying mathematical methods to novel or unfamiliar situations, and it is essential that the acquisition of higher abilities be given more importance.

In Bloom there are six categories, of which the first is knowledge and the other five are differentiated from it by describing them as levels of 'understanding'. In almost every situation there are several levels of understanding possible. For example, if an objective is left as 'the pupil understands the use of formulae' and the content list includes

$$\frac{1}{f} = \frac{1}{u} + \frac{1}{v}$$

then two levels of understanding might be

(a) the ability to apply the formula to solve a problem,
(b) the ability to understand the formula in terms of its physical meaning.

In mathematics, understanding can range from an awareness of simple relationships to the comprehension of complex systems. Since there is also a need to resolve some of the confusion in, and to establish accuracy of communication regarding educational objectives and related matters, then accuracy in precisely identifying what is meant by a particular term is essential. Without a classification of objectives in which each category is ordered from the specific to the general and from the concrete to the abstract, the danger is that understanding abilities are interpreted at low levels, with the higher level ones not considered.

LIST OF MATHEMATICS OBJECTIVES IN THE COGNITIVE DOMAIN

A (i) *Knowledge and information:* the ability to recall definitions, notations, concepts and theories.

  (ii) *Techniques and skills:* the use of straightforward calculation and computation, and the ability to manipulate symbols; solutions.

B   *Comprehension:* the ability to translate data from one form to another, for example, verbal into graphical and vice versa; to interpret or deduce the significance of data and to follow and extend reasoning; to solve problems where choice of operation is necessary.

C   *Application:* the ability to apply knowledge to novel situations presented in an unfamiliar way.

D   *Higher abilities:* this broad category covers the categories of

Analysis, Synthesis and Evaluation in Bloom's Taxonomy and includes processses such as (a) the ability to analyse given information into its various parts; (b) the ability to put together given elements to form an entirely new pattern or structure; (c) the ability to make a judgment as to the value of information as a result of analysis; (d) the ability to solve problems which involve generalization, evaluation, proof, induction, or inference and to determine the adequacy of a set of data presented for answering a problem posed.

Illustration of categories by means of test items is recommended by Bloom and others as being the best method of making a detailed and precise definition.

A (i) *Knowledge and information*

In this category the pupil is required only to recall a definition or a fact and no understanding of the knowledge is necessary. It is important to note that knowledge refers to the ability to repeat, not the ability to use. Items testing objectives in this area will be posed in exactly the same way as the material was learned.

The main sub-categories of knowledge are:

(a) *Knowledge of terminology:* The pupil is required to recognize and be familiar with the language of mathematics, i.e. the large number of terms and symbols that make up the short-hand used by mathematicians for purposes of communication. For example, the definition of technical terms such as element of a set, variable, relation, function, etc.

(b) *Knowledge of specific facts:* This objective requires the pupil to recall formulae and relationships. For example, the ability to quote the general equation of an ellipse or the formula for the circumference of a circle is indicative of this behaviour.

(c) *Knowledge of ways and means of dealing with specifics:* This sub-category includes knowledge of conventions, for example that capital letters are used to describe geometric figures, and knowledge of classifications and categories, for example whether or not a number is a member of a particular number system.

(d) *Knowledge of principles and generalizations:* This category requires the pupil firstly to recall abstractions in mathematics that help to describe, explain or predict phenomena, and secondly to recognize or recall the principles and generalizations, or specific illustrations of them, necessary in a particular problem. Knowledge of mathematical theorems and of fundamental logical principles falls into this sub-category.

The following are some examples in which the objective is knowledge. At the end of the course the pupil should be able to

(a) state the definition of an acute angle;
(b) state the theorem of Pythagoras for a plane right-angled triangle;
(c) recognize symmetry, rotation and the translation of figures in space;
(d) recall that the volume of a prism is one-third of the area of the base multiplied by the perpendicular height;
(e) define the terms mode, median and mean;
(f) recall the conventional order of operations for simplifying an arithmetic or algebraic expression;
(g) state that all equilateral triangles are similar;
(h) recognize when the precision of measurement given is of a degree justified by the nature of the question, for example, the rule for 'rounding off' numbers;
(i) recall the fundamental conditions of congruence for two triangles;
(j) recognize the limitations of inductive generalizations in proof.

Example 4.2
Items which test knowledge are:
1. The number of significant figures in 0.057 60 is ......
2. A cubic centimetre is a unit of

    A Length
    B Area
    C Volume
    D Weight

3. The fifth term of an arithmetic progression whose first term is a and whose common difference is d is given by

  A ad^5
  B $a + 5d$
  C $a + 4d$
  D ad^4
  E $a + 6d$

4. In the real number system the identity element for multiplication is ......

5. Which of the following sets of conditions is *not* sufficient for the congruence of $\triangle FGH$ and $\triangle PQR$ when f is less than g?

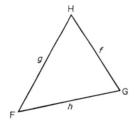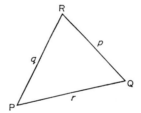

A $\hat{F} = \hat{P}$
    $g = q$
    $f = p$

B $\hat{F} = \hat{P}$
    $h = r$
    $\hat{G} = \hat{Q}$

C $g = q$
    $\hat{F} = \hat{P}$
    $h = r$

D $h = r$
    $g = q$
    $f = p$

$$E \quad \begin{aligned} f &= p \\ \widehat{G} &= \widehat{Q} \\ h &= r \end{aligned}$$

6. Which of the following is the square of an integer?

   A 71
   B 91
   C 111
   D 121
   E 141

7. The result of an operation on the numbers 9 and 8 is 17. In this operation the number 17 is the

   A Product
   B Quotient
   C Sum
   D Average
   E Difference

8. Which of the following numbers in base two is (are) even?

    I 1011011
   II 110110
  III 101001
  IV 100100

   A I only
   B I, III and IV only
   C II and IV only
   D I and III only
   E IV only

9. The only one of the following sets which is *not* a group under addition is the set of

   A Odd Integers
   B Rational Numbers
   C Real Numbers
   D Even Integers
   E Integers modulo 4.

As has been mentioned several times it is essential to have information with regard to what pupils have been taught and how they have been taught, before a particular item can be placed into one category or another. It is assumed for the last item above that the pupils have explicitly learned the material, otherwise it does not belong to the category of knowledge.

The knowledge category is indispensable to all higher categories, since the more knowledge a pupil possesses the more likely he is to succeed at these levels. However, as has been emphasized before, this category must not dominate in any list of objectives at the expense of more exacting and important higher abilities. There are several reasons for this, for example,

(a) concentration on knowledge neglects processes which can never be mastered by memorizing facts,
(b) knowledge represents a low level of mathematical performance.

Nevertheless development of knowledge is an important outcome of learning, and all the other categories assume it as a prerequisite. Moreover it is easily evaluated by objective items.

A(ii) *Techniques and Skills*

This objective includes the use of algorithms such as manipulative skills and the ability to be able to perform straightforward computations, simplifications and solutions similar to examples that the pupil has already seen in the classroom, although different in detail. The question may be such that no decision is required on how to approach the solution, only the use of a technique which has been learned, or it may be that a rule must be recalled and then a straightforward technique used, again one which has been taught.

The following are some examples in which the objective is technique. At the end of the course the pupil should be able to:

(a) find the solution sets of linear equations and inequations of the first degree in one variable;
(b) factorize expressions of the form, $ab + ac$, $a^2 - b^2$, $ax^2 + bx + c$;
(c) differentiate composite functions, for example defined by $f(x) = (2x - 5)^4$, $f(x) = \sin(ax + b)$, $f(x) = \cos^n x$;
(d) make the standard geometrical constructions using ruler,

protractor, compasses etc., for example, triangles, quadri-
laterals etc.;

(e) substitute numerical values into given formulae and evaluate
algebraic expressions;

(f) express very large or very small given numbers in the form
$a \times 10^n$ where n is a positive or negative whole number
and $1 \leqslant a < 10$;

(g) handle mathematical instruments;

(h) solve a formula for another letter or variable.

Example 4.3
Items which test technique are:

1. Construct parallelogram ABCD with AB $=$ 5 cm etc.

2. Express the numbers 1728 and 0.0027 in standard form
and find the quotient of the first divided by the second
giving the answer in standard form.

3. In the solution of the system of equations

$$2x + y = 7$$
$$x - 4y = 4$$

the value of y is equal to

    A -9
    B $-\frac{5}{3}$
    C $-\frac{1}{9}$
    D $\frac{1}{9}$
    E $\frac{5}{3}$

4. If $P = LW$ and if $P =$ 12 and $L =$ 3, then W is equal to

    A $\frac{3}{4}$
    B   3
    C   4
    D   12
    E   36

5. What are all the values of x for which the inequality
$5x + \frac{5}{3} \leqslant -2x - \frac{2}{3}$ is true?

    A $x \leqslant -\frac{7}{9}$
    B $x \leqslant -\frac{1}{3}$
    C $x \geqslant 0$

D $x \geqslant \frac{7}{3}$

E $x \geqslant \frac{9}{3}$

6. Solve the equation $\sqrt{(x + 5)} - \sqrt{(x - 3)} = \sqrt{x}$

7. The expression

$$\frac{2}{\sqrt{5}} + \frac{\sqrt{45}}{5} + \frac{1}{\sqrt{5} - 2}$$

is equal to

A $2\sqrt{5} + 2$

B $2\sqrt{5} - 2$

C 2

D $2\sqrt{5}$

E $2 - 2\sqrt{5}$

8. $\int (x - 1)^2 \, dx$ is equal to

A $2(x - 1) + k$

B $\frac{1}{2}(x - 1)^2 + k$

C $\frac{1}{3}(x - 1)^3 + k$

D $\frac{1}{3}(x^3 - x) + k$

E $(x - 1)^3/x + k$

9. The result of adding $5x/y$ and $5x/z$ is

A $\dfrac{10x}{y + z}$

B $\dfrac{5x}{yz}$

C $\dfrac{5x}{y + z}$

D $\dfrac{5x(y + z)}{yz}$

E $\dfrac{25x^2}{yz}$

10. If $f(x) = 2x + 1$ and $g(x) = 3x - 1$ then $f(g(x))$ is equal to

A $6x - 1$

B $6x + 2$
C $x - 2$
D $5x$
E $6x^2 + x - 1$

11. If $f(x) = 3x^2 + 5x - 6$, then $f'(x) =$ ....

12. If $P = \begin{pmatrix} 1 & 0 \\ 2 & 3 \end{pmatrix}$ and $Q = \begin{pmatrix} 2 & -1 \\ 0 & 1 \end{pmatrix}$ then PQ is equal to

A $\begin{pmatrix} 0 & -3 \\ 2 & 3 \end{pmatrix}$

B $\begin{pmatrix} 2 & 0 \\ 1 & 3 \end{pmatrix}$

C $\begin{pmatrix} 2 & -1 \\ 4 & 1 \end{pmatrix}$

D $\begin{pmatrix} 2 & 1 \\ 0 & 3 \end{pmatrix}$

E $\begin{pmatrix} 2 & 4 \\ & -11 \end{pmatrix}$

13.

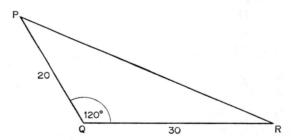

In $\triangle PQR$, PR is equal to

A $10\sqrt{13}$
B 40
C $10\sqrt{3} + 20\sqrt{2}$
D $10\sqrt{19}$
E $10\sqrt{7}$

14. If $S = 2\pi rh + 2\pi r^2$, then h is equal to

A $\dfrac{S}{2\pi r} - 2\pi r^2$

B $\dfrac{S - 2\pi r^2}{2\pi r}$

C $S - 2\pi r^2 - 2\pi r$

D $\dfrac{2\pi r^2 - S}{2\pi r}$

E $S - r$

15. If $x = 2$, $y = -1$, $z = -3$ then $y^3 - z^3 + 3xyz$ is equal to

A -10
B 24
C 34
D 44
E 46

It is perhaps relevant at this stage to mention that categories are independent of difficulty or level of mathematics. It does not follow that the higher the category, the more difficult an item is. This in fact adds another dimension to those of specific objective and content when constructing a test. It is possible to devise three levels of difficulty for technique questions, namely easy, average and hard, and similarly for the other four major categories. Hence a question defined to be hard in category A may prove more difficult for a particular group of pupils than one defined as easy in category B.

B Comprehension

This category represents the lowest level of understanding where the pupil knows and can make use of the material communicated without necessarily relating it to other material, or seeing it in all its implications. Items determine whether the pupil has grasped the meaning of the material without requiring him to apply it or analyse it.

Comprehension behaviours can be subdivided into *three* types which are hierarchical in nature:

(a) translation,
(b) interpretation,
(c) extrapolation.

Interpretation includes translation, and extrapolation both (a) and (b).

(a) *Translation*

This is the intellectual process of changing ideas in a communication into parallel forms. The pupil is required to change from one language to another or from one symbolic form to another, for example when a verbal statement of a relationship is translated into a formula or when an algebraic formula is translated into a graphical statement of the relationship. Another case of translation is to recognize or produce examples of illustrations of given definitions, statements or principles. Given recorded data, the ability to prepare graphical representations falls into this category. The thinking involved is literal and does not require the discovery of intricacies, implications or subtleties.

The following are examples of objectives in which the category is translation. At the end of the course the pupil should be able to

  (i) write the equation to represent a given graph;

 (ii) change from denary numbers to binary form and vice versa;

(iii) translate geometric concepts given in verbal form into visual or spatial form;

(iv) write in symbolic form the given verbal statement of an identity and vice versa;

 (v) change percentages into fractional terms;

(vi) change decimals into fractions and vice versa.

Example 4.4
Items which test translation are:

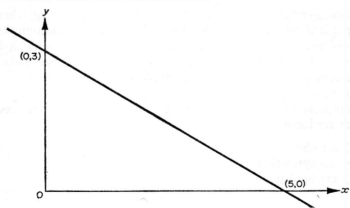

1. Write down an equation to represent the graph of the given straight line.

2. Express 11001_{two} as a denary number.

3. During his summer holidays a boy earned a weekly wage of £p for q weeks. If he spend £r altogether in that time how much money (in £) did he save?

  A $(p - r)q$
  B $(p/q) - r$
  C $qr - p$
  D $pq - r$
  E $p - qr$

4. The statement 'Of any four consecutive positive integers, the sum of the squares of the first and the last exceeds by four the sum of the squares of the middle two integers' can be expressed symbolically as:

  A $n^2 + (n + 2)^2 + 4 = (n + 1)^2 + (n + 3)^2$
  B $n^2 + (n + 3)^2 + 4 = n + 1)^2 + (n + 2)^2$
  C $[n + (n + 3)]^2 \quad = [(n + 1) + (n + 2)]^2 + 4$
  D $n^2 + (n + 3)^2 \quad = (n + 1)^2 + (n + 2)^2 + 4$
  E   none of the above

5. Which of the following represents the shaded portion in the given diagram?

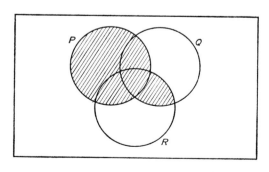

  A $(P \cap Q) \cup R$
  B $(P \cap \bar{Q}) \cap R$

C $(P \cup Q) \cap R$
D $P \cap (Q \cup R)$
E $P \cup (Q \cap R)$

(b) *Interpretation*

The essential behaviour in interpretation is the identification and comprehension of the major ideas included in a communication as well as an understanding of their interrelationships. It involves an explanation or a summarization of the communication. In items the pupil is presented with a communication, for example a graph or table of data, and asked to supply or recognize inferences which may be drawn from it. The pupil is required to show judgment by sifting the important facts from the less important and then rearranging the material to see the content of the communication as a whole.

Mathematics problems which fall into this category will be familiar in that the pupil will have seen similar types previously, but some comprehension of the underlying concepts will be necessary to solve the problem. A decision will have to be taken not only on what to do but how to do it.

The following are examples of objectives in which the category is interpretation. At the end of the course the pupil should be able to

   (i)   decide on the validity of a chain of reasoning;
  (ii)   make a deduction given a set of restraining conditions;
 (iii)   discriminate between closely related concepts, processes etc., for example be able to distinguish between different types of variation;
 (iv)   identify the operations of union and intersection of given sets;
  (v)   identify uses of the commutative, associative and distributive laws;
 (vi)   interpret charts, diagrams and tables, and identify the main points illustrated in them;
(vii)   compare related mathematical concepts, processes, figures etc.,
(viii)   see symmetry in common geometric shapes, for example isosceles triangles, equilateral triangles, rectangles etc.

Example 4.5
Items which test interpretation are:

1. The area of a certain square is represented by the expression $9x^2 + 6xy + y^2$. What will be the perimeter P of this square expressed in terms of x and y? (This is interpretation since interrelationships as well as the elements of the problem must be understood.)

2. What are the value(s) of x which satisfy the open sentence $4(6x - 1) = 3(5 + 8x)$?

   A $\frac{19}{48}$ only
   B   all values
   C   no values
   D $-\frac{19}{48}$ only
   E   all values except $\frac{1}{6}$ or $-\frac{5}{8}$

3. In the Venn diagram the numerals represent the number of elements in each area. Find $n[Q \cap (P \cup R)]$

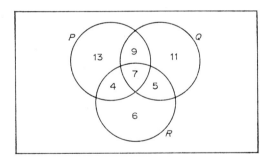

   A   16
   B   20
   C   21
   D   44
   E   55

4. The variables x and y are connected by the relationship $x^2y = k$, where k is a constant. When x is decreased in the ratio $2 : 3$ then y is

   A   increased in the ratio $9 : 4$

B   decreased in the ratio $4:9$
C   unchanged
D   increased in the ratio $3:2$
E   decreased in the ratio $2:3$

5. If the two lines $8x - 12y - 3 = 0$ and $px + 2y - 7 = 0$ are perpendicular then p is equal to

A -3
B $-\frac{4}{3}$
C $\frac{1}{3}$
D $\frac{3}{4}$
E 3

6. Which of the following fractions is between $\frac{3}{4}$ and $\frac{5}{6}$?

A $\frac{2}{3}$
B $\frac{17}{24}$
C $\frac{19}{24}$
D $\frac{21}{24}$
E $\frac{11}{12}$

7. There are many examples in mathematics in which the pupil is required to display the ability to interpret graphs and tables. For example, in an item where a pie chart, a line graph, a bar graph or a table of statistics is given, the pupil can be asked to demonstrate this behaviour.

(a) Use the graph at the top of page 59 to answer the following questions.

   (i) Three hours after starting, car A is how many kilometres ahead of car B?

   A   2
   B   10
   C   15
   D   20
   E   25

   (ii) How much longer does it take car B to go 50 kilometres than it does for car A to go 50 kilometres?

   A   1 hour 15 minutes
   B   1 hour 30 minutes

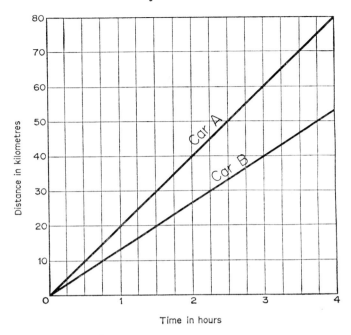

C   2 hours
D   2 hours 30 minutes
E   2 hours 35 minutes

(b) The diagram inside the rectangle shows parts of two
graphs, one a straight line, $y = g(x)$, and the other a
'wave' curve, $y = f(x)$. The straight line *touches* the curve
at Q and U.

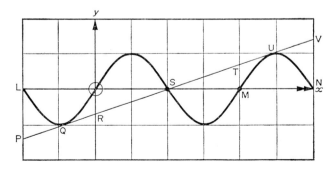

(i) Between which of the points L, O, S, M, N is $f(x) \leqslant 0$ when $x \leqslant 0$?

A   S and M only
B   L and O only
C   L and M only
D   This never happens.

(ii) Between which of the points, P, Q, R, S, T, U, V is $g(x) \geqslant f(x)$?

A   Q and S only
B   S and U only
C   S and V only
D   T and V only

(iii) Between which points on the *straight line* is $g(x) \geqslant 0$ when $f(x) \leqslant 0$?

A   S and T only
B   S and U only
C   T and V only
D   U and V only

(iv) Between which points *on the curve* is $f(x) \leqslant 0$ when $g(x) \leqslant 0$?

A   This never happens
B   L and S only
C   S and M only
D   L and O only

(v) At which point(s) is $f(x) = g(x)$?

A   Only at S
B   Only at Q and U
C   Only at R
D   At Q, S and U

8. ABCDE is part of a figure. PQ is an axis of symmetry for the complete figure. Copy and sketch in the other part which completes the full symmetrical figure.

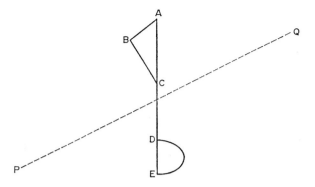

(c) *Extrapolation*

This objective involves the ability of the pupil to extrapolate or extend trends or tendencies beyond given data. *Interpolation* is regarded for the purposes of classification as a type of extrapolation in that judgments with respect to intervals within a sequence of data are similar to judgments going beyond the data and in either case the underlying principle must initially be grasped. There must be an awareness of the limits of the data and also possible limits within which it can be extended. Any inference made will have some degree of probability. Extrapolation is an extension of interpretation in that once the pupil has interpreted the material, he is required to specify any implications, consequences or effects of it, that is, once the data has been encoded then an extrapolation item requires the pupil to utilize the code to predict outcomes.

The following are examples of objectives in which the category is extrapolation. At the end of the course the pupil should be able to

(i) interpolate where there are gaps in data, for example, in a given graph;

(ii) predict population characteristics of sample data, for example, given a graph of the average weekly rainfall in Britain last year he is able to predict when and where rainfall is likely to be the highest next year. In recognizing a pattern he is translating and interpreting the data and in predicting he is going beyond what is given and reaching another stage of understanding;

(iii) infer the degree of a function given its graph;

(iv) complete gaps and continue the sequence in given number patterns;

 (v) extend the ideas present in one situation to another relevant situation.

Example 4.6
Items which test extrapolation are:

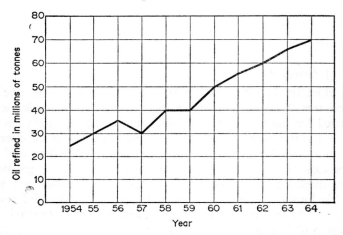

1. Is it possible to interpolate from the given data? Can an estimate be made of the oil likely to be refined in 1965? If so, what would your estimate be?

   In graph questions it is often easier to measure outcomes in interpretation and extrapolation together, by basing a series of items on a given situation. Questions of this type are easy to find and several interesting examples are to be found in a book by Morse and McCune.

2. Which of the following numerals represents the largest number?

   A 1011_{two}
   B 102_{three}
   C 23_{four}
   D 21_{five}
   E 20_{six}

3. Which of the following formulae defines the function represented by the given graph?

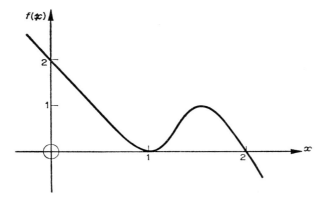

A $f(x) = (1 - x)(2 - x)$
B $f(x) = (1 - x)(x - 2)$
C $f(x) = (1 - x)(2 - x)^2$
D $f(x) = (1 - x)^2(x - 2)$
E $f(x) = (1 - x)^2(2 - x)$

4. Find the missing term and the next terms in the following sequence of base five numbers, expressing the answer in base five.

1, 4, 14, 31, ..., 121, 144, ..., ...

C. *Application*

This category refers to the use of general ideas, principles or methods to new situations. The items require the pupil to apply familiar concepts to unfamiliar situations, that is, to apply knowledge and understanding of the skills to new situations or to situations presented in a novel manner, sometimes referred to as 'transfer of training'. The method of solution is not implied in the question, and the ability sought after is that of being able to develop the steps in the solution of the mathematical problem, rather than that of being able to reproduce a classroom solution. The thinking process involved is higher than that of comprehension, due to the unfamiliarity and problematic nature of the given situation. It is essential that the situations presented to the pupil are different from those in which he originally learned the meaning of the abstractions that he will be required to apply, to ensure that the problem cannot be solved by routine methods. The category is necessary since the comprehension of an abstraction does not guarantee that the pupil will be able to recognize its relevance and apply it correctly in real life situations. The ability to be able to apply acquired concepts and principles to a new problem or to select the correct abstraction for one that appears unfamiliar until the elements have been restructured into a familiar context, is of extreme importance in all mathematics courses since most of what a pupil learns is intended for application to everyday problem situations.

The following are examples of objectives in which the category is application. At the end of the course the pupil should in general be able to use the given ideas, principles and methods to solve problems described in the form of situations in everyday life and examples from other branches of mathematics. For example, the pupil should be able to

(a) apply the laws of trigonometry to practical situations and surveying problems;

(b) apply the knowledge he has of the quadratic formula and the ability he has to be able to state it in his own words, to solve a quadratic equation that he has selected to generate the solution of a new problem;

(c) apply differentiation to problems involving maxima and minima;

(d) apply a mathematical model to solve a practical problem, for example linear programming problems;

(e) select the most appropriate formulae, method or process to solve a problem.

It is difficult to give examples which will always be tests of application, since items which represent the category under review for some teachers might involve only knowledge or comprehension for pupils of teachers who have studied the specific problem presented in the item, in the class textbook or in classroom discussions. Bloom suggests three approaches in an attempt to set up a new situation:

(i) present a fictional situation,

(ii) use material with which pupils are not likely to have had

contact, such as simplified versions of complex problems studied in more advanced work, and
(iii) take a new slant on common situations.

The third of these would appear to be the most suitable source of new yet realistic items, but it is difficult to find unfamiliar information at an appropriate level to use as the basis for a question. The solution in an application item need not be difficult but the pupil is required to demonstrate some insight in finding it.

A traditional essay-type mathematics problem which requires the pupil first of all to exhibit a type of comprehension, for example translation or interpretation, followed by the selection and use of one or more previously learned algorithms, will belong to this category. Similarly if a problem requires the pupil to choose, use and combine several principles or algorithms in an unfamiliar way then it is testing application. In both cases it is necessary for the pupil not to have been exposed to a problem of the same form in the classroom, or the category will be on the level of comprehension or even knowledge. It is also vital to ensure that the problems cannot be solved on the basis of general knowledge only, since application is concerned with the extent to which the student has learned to apply the facts, principles and procedures taught in a course.

Example 4.7
Items which test application are:

1. A circle is inscribed in a triangle XYZ, touching XY at P; if the length of XY is 7, of YZ is 6, and of ZX is 8, what is the length of XP?

   A $3\frac{1}{2}$
   B 4
   C $4\frac{1}{2}$
   D $4\frac{2}{3}$
   E 5

2. The length l cm of a rectangle is increased by p per cent and the breadth, b cm, is decreased by p per cent. The new area expressed in terms of p, l and b is

   A $lb - \dfrac{p^2}{100^2}$

B $lb\left(1 + \dfrac{p^2}{100^2}\right)$

C $lb\left(1 + \dfrac{p}{100}\right)^2$

D $lb\left(1 - \dfrac{p^2}{100^2}\right)$

E $lb + \dfrac{p^2}{100^2}$

3. Four towns P, Q, R and S are on a straight road in the order named. The distance from P to Q is three-quarters of the distance from P to S, and the distance from P to R is four-fifths of the distance from P to S. The distance from P to Q is what part of the distance from P to R?

   A $\frac{1}{2}$
   B $\frac{3}{5}$
   C $\frac{15}{16}$
   D $\frac{16}{15}$
   E $\frac{5}{3}$

This question involves comprehension possibly via a diagram followed by the application of the concept of ratio.

4. If x and y are two prime numbers and each is greater than 7, which of the following is true?

   A $x \times y$ is a prime number
   B $x - y$ is a prime number
   C $x \div y$ is a whole number
   D $x + y$ is an odd number
   E $x \times y$ is an odd number

5. The unshaded part of the figure at the top of page 67 shows a window made up of two parts, one circular and the other rectangular (XYRS). What is the ratio of the area of the circular part to the area of the rectangular part?

   A $2\pi : 3$
   B $\pi : 10$
   C $\pi : 6$
   D $\pi : 5$
   E   It cannot be determined from the given data

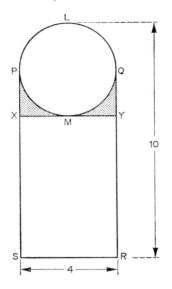

6. If $x > 1$, which of the following increase(s) as x increases?

I $x - \dfrac{1}{x}$ II $\dfrac{1}{x^2 - x}$ III $4x^3 - 2x^2$

A   I only
B   II only
C   III only
D   I and III only
E   I, II and III

In this example two principles must be understood and applied.

7. Let a binary operation $*$ on ordered pairs of integers be defined as $(a),\ b) * (c,\ d) = (a - c,\ b + d)$. Then if $(3, 2) * (0, 0)$ and $(x, y) * (3, 2)$ represent identical pairs, x equals

A -3
B   0
C   2
D   3
E   6

8. For what positive value of k will the triangle formed by the co-ordinate axes and the line $2x + ky = 6$ have area k?

9. For what values of x is $(x^2 - 2x + 3)/2$ an integer?

   A All integers except zero
   B All integers greater than 2
   C All even integers except zero
   D All odd integers
   E None of these

   The ability to comprehend the situation and to develop a series of orderly steps is the requirement of this problem.

10. On the average a boy gets $\frac{3}{4}$ of his multiplication 'sums' right and $\frac{1}{2}$ of his division 'sums' right. What is the probability that he will get $(2.7 \times 4.5)/7.2$ wrong?

11. A possible type of item is one which consists of (i) a statement of a problem situation and (ii) a list of statements or procedures which might (or might not), if used, yield information useful in solving the problem. The pupil's task is to identify the procedure which would lead to the solution.
    Item. In a square of side x cm, one of the sides is increased by 6 cm and the other decreased by 3 cm. If the area is increased by 6 cm², which of the following statements describes this situation?

    A $x^2 - (x^2 + 3x - 18) = 6$
    B $(x + 6)(x - 3)\qquad = 6$
    C $2x^2 - 3x + 18\qquad = 6$
    D $(x^2 + 3x - 18) - x^2 = 6$
    E None of these statements

D. *Higher Abilities*

This is a very broad category and includes the categories of Analysis, Synthesis and Evaluation.

As a preliminary step to certain problem-solving procedures or to the making of judgments which are based on the outcome of the solution, it is very often essential to make an *analysis* of the situation. This may take the form of, for example,

(i) breaking down the information into its relevant parts and reorganizing it in terms of relationships within a problem or configuration,

(ii) distinguishing facts from assumptions and ascertaining what assumptions might have to be made to justify certain procedures, or

(iii) checking the consistency of hypotheses with given information and assumptions.

After a careful analysis of a problem a pupil may be required to put together the elements or parts in order to form a pattern or structure not clearly visible before, for example to design an experimental procedure to solve a certain problem or to present conclusions with logically organized supporting evidence. This ability, which most clearly provides for creative behaviour and originality on the part of the pupil, is called *synthesis*. To be creative with mathematics requires the pupil to make discoveries which are original *to him*, for example he begins with some basic properties or other symbolic representation and he deduces other properties or relations, or in solving problems he demonstrates ingenuity, improvisation or the creation of new assumptions, since the elements in the problem will not restructure to take a familiar form. The principles which the pupil must be able to recognize and apply may initially seem unrelated and do not appear until an analysis of the information and its internal relationships takes place.

After analysing a problem a pupil may be required to make a judgment as to the values of the information resulting from the analysis. This ability to identify standards or values for an idea or an object and then to make value judgments is called *evaluation*. An example of this sub-category arises when a pupil is given a new piece of work and is asked to locate errors within it.

Questions on any of the above three categories will fall into the broader category of higher abilities.

There will always be difficulties in deciding whether to assign a mathematical problem to Application or to Higher Abilities. The basic difference is essentially that between a pupil being able to *reproduce* well-comprehended rules and procedures in order to solve unfamiliar problems, and being able to *produce* something that is entirely new to him by discovering relationships among previously unrelated principles and algorithms when no straightforward method is available to provide the entire solution. Another difficulty is that the solution to a problem can often be arrived at by two different methods one of which represents thinking at category D level but the other, usually a time-consuming application of a known process, represents thinking at a lower level.

The following are examples of objectives in which the category is higher abilities. At the end of the course the pupil should be able to

(a) analyse information into its underlying parts and establish a correct relationship between them;

(b) distinguish a conclusion from statements which support it;

(c) detect logical fallacies in arguments or judge the reasonableness of answers in problems involving statistical inference;

(d) proceed from hypothesis to a conclusion, for example analyse a newspaper statement and determine what is assumed and whether the suggested conclusions follow from the given facts or assumptions;

(e) make mathematical discoveries and generalizations from a variety of results;

(f) construct a proof or a problem new to him;

(g) reason creatively in mathematics;

(h) invent a new mathematical operation or structure;

(i) produce a plan or develop a procedure, for example develop a layout for a sports field;

(j) abstract, symbolize and generalize (in the same problem);

(k) indicate logical fallacies in arguments;

(l) judge the significance of a problem;

(m) validate answers or judge the correctness of a proof by an internal analysis of the steps;

(n) solve problems which involve generalization, induction, proof, inference or sufficiency of data.

Example 4.8
Items which test the higher abilities are:

1. Let m and n be any two odd numbers with n less than m.

The largest integer which divides all possible numbers of the form $m^2 - n^2$ is

A 2
B 4
C 6
D 8
E 16

In this problem the pupil is required to break the problem down into parts and recall knowledge and comprehension about each part. It is unlikely to be restructured into a familiar form and then solved, which would put it into the category of application.

2. Given three positive integers a, b and c, with greatest common divisor D and least common multiple M, which two of the following statements are true?

   I the product MD cannot be less than abc.
  II the product MD cannot be greater than abc.
 III MD equals abc if and only if a, b, c are each prime.
 IV MD equals abc if and only if a, b, c are relatively prime in pairs. (That is, no two have a common factor greater than 1.)

A I and II only
B I and III only
C I and IV only
D II and III only
E II and IV only

3. One of the few examples on Mathematics from Bloom's Taxonomy.
Statement of facts. The following table represents the relationship between the yearly income of certain families and the medical attention they receive.

Family income	Per cent of family members who received no medical attention during the year
Under $1200	47
$1200 to $3000	40
$3000 to $5000	33
$5000 to $10000	24
Over $10000	14

Conclusion: Members of families with small incomes are healthier than members of families with large incomes.

Which one of the following assumptions would be necessary to justify the conclusion?

A   Wealthy families had more money to spend for medical care.

B   All members of families who needed medical attention received it.

C   Many members of families with low incomes were not able to pay their doctor bills.

D   Members of families with low incomes often did not receive medical attention.

This is an example of items that require to be constructed to test the abilities to

(a) distinguish between facts that are relevant or irrelevant to an argument,

(b) identify those basic assumptions upon which the proof or conclusion depends.

4. Assuming that $\sqrt{2}$ is an irrational number, we can prove that $a + \sqrt{2}$, for any rational a, must also be an irrational number. Which one of the following statements can be used to obtain such a proof?

A   The difference between two rational numbers is always a rational number.

B $\sqrt{3}$ is also an irrational number.

C   The sum of two irrational numbers is sometimes rational and sometimes irrational.

D   The product of a rational number, different from zero, by an irrational number is always an irrational number.

E   None of the above can be used to prove the theorem.

5. Invent a binary operation which combines elements of the form $\begin{bmatrix} a \\ b \end{bmatrix}$ where a and b are integers. Investigate some of the properties of your operation and see if the set of these elements forms a group under your operation.

Since synthesis places emphasis on creativity, objective test items would appear to have a limited application in this area and the ability is better tested by other forms of assessment, for example open ended or free response items. However, multiple choice questions, if carefully constructed, especially with regard to the alternatives, can assess this level with a certain amount of success, although this type of item should not be the teacher's only means of measuring a pupil's attainment of the higher abilities.

6. Consider a number system based on the symbols X, I, $\bar{X}$, Θ, IX, II, ..., which correspond to 0, 1, 2, 3, 4, 5, ...

   (a) Write the next three numbers in the above sequence of new symbols
   (b) Write the numerals for 53 and 89
   (c) What numbers are represented by the numerals $\bar{X}X$ and ΘIX?

   Find:
   (d) the sum $I\Theta X + IX\bar{X}$
   (e) the difference $\Theta X - IX$
   (f) the product $XI\Theta \times \bar{X}I$
   (g) the quotient $\bar{X}X \div \bar{X}$

7. It has been proved that an infinite number of triangles possess a property Q.
   Statement S: All triangles possess property Q.
   Which of the following is necessarily correct?

   A S is true, and no further proof is required.
   B S is true, but proof is required.
   C S is more likely to be true than false.
   D S is more likely to be false than true.
   E None of the above.

8. In the following series each item after the first two is the sum of the previous two terms. What is the ratio of the number of even to the number of odd items in the first ninety terms?

I   I   2   3   5   8   13   .   .   .   .   .

A $\frac{20}{70}$

B $\frac{30}{90}$

C $\frac{30}{60}$

D $\frac{40}{50}$

E $\frac{45}{45}$

9. In the evaluation of $\int_{-9}^{-5} dx/(x+1)$ a pupil wrote down the following steps.

I $\int_{-9}^{-5} dx/(x+1) = \left[\log_e (x+1) \right]_{-9}^{-5}$

II $\left[\log_e(x+1) \right]_{-9}^{-5} = \log_e(-4) - \log_e(-8)$

III $\log_e(-4) - \log_e(-8) = \log_e(-4/-8)$

IV $\log_e(-4/-8) = -\log_e 2$

In which step (if any) does the *first* mistake in the evaluation occur?

A   I

B   II

C   III

D   IV

E   None of these

10. Investigate winning strategies at noughts and crosses and generalize your results.

11. By simplification, the equation

$$\frac{2x^2}{x-1} - 4 = \frac{6x-4}{x-1}$$

can be reduced to the equation $x^2 - 5x + 4 = 0$. The roots of the latter equation are 4 and 1; then the root(s) of the first equation is (are):

A   4 and 1

B   Neither 4 nor 1

C   Only 4

D   4 and some other root

E   Only 1

12. Examples of the type where two or more premises are given and the choice of a valid conclusion has to be made, or given a problem which has to put into the form of categorical syllogism and then the truth of the conclusion proved or disproved, fall into this category.

   Consider the statement: Studying hard is a necessary condition for getting an A. This is equivalent to which of the following:

   I   If you study hard, then you will get an A.
   II   If you get an A, then you have studied hard.
   III If you don't study hard, then you will not get an A.
   IV   If you don't get an A, then you have not studied hard.

   A   I only
   B   I and II only
   C   II only
   D   II and III only
   E   I, II, III and IV

13. Compare the assumptions and development of Euclidean geometry with those of transformation geometry and defend the statement that the latter has been successful in replacing the former.

14. All 'data sufficiency' items will be available for this category and examples of this type of item, which requires the pupil to reason logically in considering whether the data given is relevant to a solution, will be discussed in a later chapter.

MATHEMATICS TEACHING OBJECTIVES IN OTHER DOMAINS

Two other domains, the affective and the psychomotor, are considered to have objectives which should be considered by the teacher in any scheme of mathematical achievement.

Krathwohl has pointed out the difficulties experienced by the authors in formulating the hierarchical classification in the affective domain. The common thread running through the processes is the concept of 'internalization' which 'refers to the inner growth that occurs as the individual becomes aware of and then adopts attitudes, principles, codes and sanctions which become inherent in forming value judgments and in guiding his conduct'. The point that nearly all cognitive objectives have an affective component has been made, since most teachers hope that their pupils will develop a continuing interest in the mathematics taught. Although they may be implied, it is usual to find objectives which involve attitudes towards, interests in and appreciations of mathematics left unspecified. In certain cases, however, if the affective objective refers to the whole course then it would be better to emphasize it by listing it as a separate behaviour. As a measurable change in this behaviour might not be attained until near the end of a pupil's school career, there is an even greater need for specification.

At present several courses mention in the preamble which usually precedes the topics in the syllabus, such statements as 'It is desirable to encourage in pupils at all levels an enthusiasm for mathematics' or 'Goals should include the creation and maintenance of interest in, and regard for, the subject.' The development of positive attitudes towards mathematics is important for several reasons, for example if while learning mathematics the pupil acquires a dislike for it then further learning is improbable and part of the instructional process is lost.

A list of affective factors, some of which might be included in a course of mathematics is as follows: at the end of the course the pupil will have acquired

(a) an awareness of the influence, contribution, relation and value of mathematics to other subjects in the curriculum;

(b) an awareness of the contribution, value and role of mathematics in society;

(c) an awareness of the beauty of geometric shapes in the environment;

(d) an awareness of the power of mathematics in general;

(e) a willingness to respond, to ask questions, to take part in discussions and to express his own point of view;

(f) a willingness to work co-operatively with other members of the class in mathematical activities;

(g) sufficient involvement with mathematics to gain satisfaction in working with it;

(h) a desire to make discoveries, to concentrate and to assume responsibility for an assignment;

(i) an interest and willingness to participate in mathematical activity in his leisure time—this is observable by membership of a mathematics society or a computer club;

(j) an interest and enjoyment in mathematics which will encourage him to continue in the study of the subject—this is indicated by the pupil selecting mathematics when options are offered at various stages in school, or by choosing to study the subject at university;

(k) a desire to read literature on the subject arising from the enjoyment that this activity gives him—this is observable by the use the pupil makes of the books in the school mathematics library;

(l) habits of persistence, participation, conscientiousness, thoroughness etc.;

(m) the development of study habits essential for independent progress in mathematics;

(n) attitudes that lead to self-understanding, self-disciplining, self-respect, initiative, independence and the development of a self-critical awareness.

It is difficult to measure objectives in the affective domain and assessment instruments in this field are not as well developed as those for the testing of the attainment of cognitive objectives. Attitude assessment tests, questionnaires, essays and self-report procedures are possible means of measuring the attainment of objectives, in addition to the observational methods mentioned in listing the above behaviours. This assessment would require to be of the continuous type over the four years of an 'O' grade course, for example, and would give a record of the extent to which a pupil has acquired various behaviours.

The outline of a taxonomy of educational objectives in the psychomotor domain is a tentative hypothesis to fill the gap in a tripartite structure which is necessary to obtain a comprehensive scheme of educational objectives. It is difficult to visualize many objectives of this third type in a mathematics course, apart from the development of skill

(a) in using hand-operated calculating machines;

    (b) in using mathematical instruments, for example, ruler and compasses;

    (c) in using drawing, both freehand and precise, for example to construct triangles and to solve trigonometry problems;

    (d) in sketching geometric figures.

CONCLUSION

One obvious conclusion is that the teacher who makes use of educational objectives, either as a basis for examination construction or course construction, has a decided advantage over one who either chooses to ignore them or is unaware of recent developments in this field. The pupil must also benefit from this approach, since he normally directs his studies according to what he thinks the teacher's objectives are and if these are made clear to him in the form of precise statements of behaviour changes intended to be brought about by what he learns, then he is not only likely to co-operate but he is in a better position to do so.

It would appear that in the assessment of the attainment of objectives the emphasis will continue to be placed on testing cognitive behaviour changes. However, if attitude objectives are accepted as appropriate for school mathematics then greater provision for their testing must be made. This will require the development of some suitable test instruments to measure such attitudes and there is a great deal of research yet to be done in this field. The active participation by teachers in defining objectives and developing tests can only lead to an improvement in examinations in general, curriculum development and the learning process as a whole.

5. Types of Objective Test Items

Objective test items can be classified as either fixed-response or completion.

COMPLETION ITEMS

The following are examples of completion items.

Example 5.1
What is the image of (3, 4) under the dilation [O, −2] where O is the origin?

In this item the candidate has to supply the answer. There is at first sight greater scope for this type in mathematics than there is in other subjects, since in mathematics there are more items to which there is an absolutely correct answer. The greatest disadvantage of this type of question is that it is not suitable for machine scoring, and the manual scoring of large numbers of such items can produce unacceptably high rates of error unless strict safeguards are introduced in the marking process. In using such items it is necessary to ensure that the question is so phrased that there is only one possible correct answer, and that the candidate cannot be penalized because of faulty wording by the examiner. Consider Example 5.2.

Example 5.2
Express $(x + 2y)^2$ as a sum of terms.

If the examiner sets this question he should be prepared to accept as possible answers

$$x^2 + 4xy + 4y^2$$
$$x^2 + 2xy + 2xy + 4y^2$$
$$x(x + 2y) + 2y(x + 2y)$$

Each of these answers the question as stated. It is not reasonable to expect a candidate to interpret the examiner's thought process.

It is also desirable that a completion item should ask for one piece of information only. Thus, in the dilatation question given above the candidate has to provide both an x- and a y-co-ordinate. A decision has to be made by the examiner whether to accept partly correct answers, e.g. $(-6, 8)$ or $(6, -8)$. Is this answer in some sense better than $(-9, -12)$? Whenever the possibility of part credits arises in a question there may always be borderline cases which will have to be left to the discretion of the marker, and thus one of the advantages of objective testing will be lost. On a national scale, the advantages of fixed response items are so great— on a purely administrative basis—that it would be necessary to prove not only that free response is as good as fixed response, but that it is very much better before a decision to retain such items would be made.

FIXED RESPONSE ITEMS

A fixed response item is one where the candidate is allowed to select only from a fixed number of predetermined responses. There are a great many possible formats for such items and examples quoted will illustrate some possible types of such question in mathematics.

True/False Items

A common form of fixed response item is the true–false item. In this a candidate has to state whether a given statement is true or false.

> *Example 5.3*
> If P, Q and R are sets then $P \cap Q = P \cap R \Rightarrow Q \supset R$

Such questions now tend to be looked on with disfavour, since the problem of what steps to take to deal with guessing will give rise to marking procedures to which exception can easily be taken. This problem of correction factors will be considered later (Chapter 8). Apart from this, however, there is some reason to suggest that candidates tend to give this type of item less thought than is desirable. The fact of having to make only a two-choice decision does provide a situation where guessing is encouraged. Moreover, such items tend to test only the lowest abilities and are very often marked by an attempt by the examiner to entice the candidate into a wrong choice, as in example 5.3 or even more strongly in the following example.

Example 5.4
$x^2 < 9 \Rightarrow x < 3$

Multiple Choice Items

The simplest form of fixed response item apart from the true–false item is the multiple choice item. Most examples met with in tests will be of this type. Here the candidate is given a problem and supplied with a number (usually four or five) of different responses from which he has to select the most appropriate one.

Example 5.5
For all x and y, $(x - 3y)^2$ is equal to

A $x^2 - 9y^2$
B $x^2 + 9y^2$
C $x^2 - 3xy + 9y^2$
D $x^2 - 6xy + 9y^2$
E $x^2 - 6xy - 9y^2$

Example 5.6

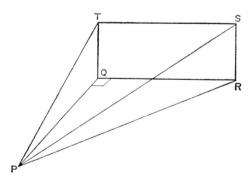

PQR is a horizontal right-angled triangle. QT and RS are equal vertical lines. Which one of the following statements is true about the sizes of angles TPQ and SPR?

A   angle TPQ $=$ angle SPR
B   angle TPQ $<$ angle SPR
C   angle TPQ $>$ angle SPR
D   The information given is insufficient to determine which of A, B or C is true.

Example 5.7
If $P = \{1, 2, 3, 4, 5\}$ and $Q = \{3, 4, 5, 6, 7\}$ are subsets of $\mathscr{E} = \{1, 2, 3, 4, 5, 6, 7, 8, 9\}$, $P \cap Q'$ equals

A $\{1, 2\}$
B $\{6, 7\}$
C $\{8, 9\}$
D $\{3, 4, 5\}$
E $\{1, 2, 3, 4, 5, 8, 9\}$

There are a few commonly used terms in connection with this type of item which may be mentioned here, although the construction of this type of item will be considered in greater detail in the next chapter. The *stem* of the item is the first part; that is, in Example 5.6 above, the sketch and the initial wording down to the end of the third sentence. The *responses* are the suggested answers —the correct response is termed the *key* and the incorrect responses are *distractors*. The stem may be in the form of an incomplete statement as in Example 5.5 or in the form of a question as in Example 5.6.

MULTIPLE COMPLETION ITEMS

In the three examples given above there was, of course, only one correct answer in each case. It may be desirable to construct an item to which there is more than one correct answer. Thus it may be required that a candidate should be able to select a number of factors of a given expression. In this case a multiple completion format may be used, as in Example 5.8.

Example 5.8
In the following questions, one or more of the given responses may be true. Answer

A   if 1, 2 and 3 only are true
B   if 1 and 3 only are true

C   if 2 and 4 only are true
D   if 4 only is true
E   if some other response or combination of responses is true

(a) Which of the following are factors of $x^3 - 6x^2 + 11x - 6$?

1. $x - 1$ 2. $x + 1$ 3. $x - 2$ 4. $x + 2$

(b) Which of the following are measures of the spread of a set of marks?

1. median   2. mode   3. range   4. standard deviation

The format of the responses is such that a knowledge of the truth or falsity of any one part still leaves a choice of three possible answers. This form of item may be open to the criticism that the response E—which must be at some time the correct response if it is to function—is really the combination of 11 or 12 different responses. There will be a fairly high chance of its being chosen for the wrong reason. However, this disadvantage would seem to be outweighed by the advantage of not giving the candidate the additional knowledge that only one of the suggested responses is true. The multiple completion format can also be used where only three completions to the stem are given. In this case a suitable format for the question would be as in Example 5.9.

Example 5.9
In the following questions one or more of the given responses may be true. Answer

A   if 1 only is true
B   if 2 only is true
C   if 3 only is true
D   if 1, 2 and 3 are true
E   if some other combination of the responses is true

(a) Which of the following expressions is/are differentiable at the point with x-co-ordinate o?

1. $x^{1/2}$ 2. x^2 3. 7

(b) Which of the following points lie(s) on the curve with equation $y = x/(x - 1)$?

1. $(0, 0)$ 2. $(1, 0)$ 3. $(2, 1)$

SITUATION SET

It is sometimes convenient to group together a number of items, each dealing with a different aspect of a particular situation. Such a combination is known as a situation set, and has the advantage that it is possible to devote more space or time to establishing the framework for a question than would be justified if only one item were to be set. In mathematics this is a particularly useful type of question in dealing with graphical situations, but can also be used where it is thought desirable to test mathematical reasoning in situations which are not explicitly covered by the syllabus. Thus it would be possible to provide a set of items each dealing with one aspect of mathematics involving modulus.

Examples 5.10–5.14 deal with a quantity called the modulus of a number. The modulus of x is written $|x|$. Where x is a real number, $|x|$ denotes the numerical value of x. Thus $|-3.5| = 3.5$, $|2.6| = 2.6$, $|-7| = 7$. Formally $|x| = x$ if $x \geqslant 0$, $|x| = -x$ if $x \leqslant 0$.

Example 5.10
$|3| + |-2| + |-1|$ equals

A   0
B   2
C   4
D   6
E   none of these

Example 5.11
$\{x: |x - 2| < 3, x \in R\}$ equals

A $\{x: x > -5, x \in R\}$
B $\{x: x < 1, x \in R\}$
C $\{x: x < 5, x \in R\}$
D $\{x: -1 < x < 5, x \in R\}$
E   none of these

Example 5.12
Consider the two expressions $|x^2|$ and $|x|^2$ where $x \in R$. These expressions are equal for

A   no value of x
B   one value of x only
C   two values of x only
D   three values of x only
E   all values of x

Example 5.13
Which of the following graphs is most likely to be the graph of $y = x^2 - |x|$?

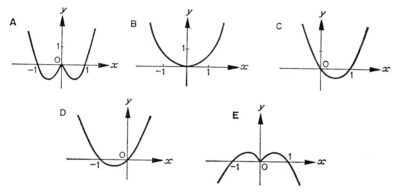

Example 5.14
The graph of $|x| + |y| = 1$, $x \in R$ is drawn. The graph will be

A   a straight line
B   the boundary of a triangle
C   the boundary of a square
D   the circumference of a circle
E   none of these

The same format can be used to test transformations, see diagram top of next page.

Example 5.15 is concerned with transformations applied to the figures PQRS, a square with vertices $(1, 1), (2, 1), (2, 2)$ and $(1, 2)$. Figures A–E on page 86 show the image $P'Q'R'S'$ of PQRS under certain transformations.

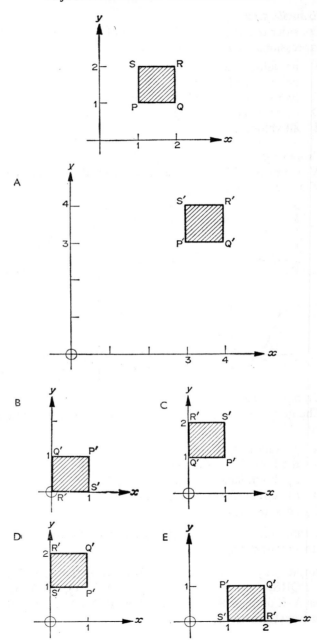

Example 5.15
In which figure is P'Q'R'S' the image of PQRS under:
(a) reflection in the line $y = 1$,
(b) an anti-clockwise rotation about $(1, 1)$ through $90°$,
(c) the dilatation $[P, -1]$,
(d) the translation $(2, 2)$?

RELATIONSHIP ANALYSIS ITEMS

Example 5.4 is one case of a fairly wide range of questions in mathematics in which candidates are required to distinguish between various possible ways in which two statements can be related. This type of question may be termed relationship analysis and could be used with the format in the following example.

Example 5.16
In each case answer

A if (1) implies (2) but (2) does not imply (1)
B if (2) implies (1) but (1) does not imply (2)
C if (1) is equivalent to (2)
D if (1) denies (2) or (2) denies (1)
E if none of the above relationships hold

(a) $x \in R$ and (1) $x < 3$ (2) $x^2 < 9$
(b) (1) PQRS is a square   (2) PQRS is a quadrilateral with rotational symmetry (cyclic symmetry) of order 4
(c) $x \in R$ and (1) $x^2 \neq 9$ (2) $x \neq 3$
(d) (1) $f'(a) = 0$ (2) the graph of $f(x)$ has a turning point where $x = a$
(e) $x \in R$ and (1) $x^2 = 9$ (2) $x = 3$
(f) $x, y \in R$ and (1) $x^2 + y^2 = 25$ (2) $x + y = 5$

Such questions are probably bound to test the highest category of mathematical thinking and are probably not suitable for immature candidates, but there might well be a place for them at a higher level.

DATA SUFFICIENCY ITEMS

A further type of mathematical reasoning which can be tested by an objective item—and which has been neglected in conventional examinations—is the ability of the candidate to decide how much information is required in order to solve a problem. The problem may be very simple—say the construction of a quadrilateral—or more complicated—say the conventional time and distance problem which has featured so often in past conventional examinations. The format used here is the data sufficiency item. The preamable again can be standardized to cover a number of items.

Example 5.17
In the following questions you are given a stem and two additional pieces of information labelled (1) and (2). In order to answer the question posed in the stem you may require to use neither, one or both pieces of information. It may be that even using both pieces of information you still do not have sufficient data to solve the problem. You are not required to solve the problem, only to decide how much information is required. Answer

A   if the information in the stem is sufficient by itself
B   if fact (1) is necessary and sufficient to solve the problem
C   if fact (2) is necessary and sufficient to solve the problem
D   if facts (1) and (2) are together necessary and sufficient to solve the problem
E   if all the information given is insufficient to solve the problem

(a) It is required to calculate the length of the diagonal PR of a quadrilateral PQRS in which PQ, PS and QS are given.
(1) the length of RS is given
(2) the length of QR is given

(b) It is required to calculate the size of angle PQR in an isosceles triangle PQR in which PQ = QR.
(1) angle QRP = 50°   (2) PQ is of length 5 cm

(c) A runner P runs twice round a running track at uniform speed. Runner Q starts at the same time as P and when Q has completed one lap at uniform speed, runner R runs the second lap at uniform speed, P and R finish the race at the same time. You should ignore the time taken for the change-over from Q to R. Given that P runs at a speed

2 km/h greater than Q, it is required to calculate P's speed in km/h.

(1) R runs at a speed 5 km/h greater than Q.

(2) The track measures 400 m in circumference.

(d) The heights of a group of children are normally distributed with a mean of 150 cm. It is required to find the percentage of children in the group with a height greater than 185 cm.

(1) there are 675 children in the group

(2) the standard deviation of the heights is 15 cm

It may be argued that this form of question asks the candidate to do too much work in absorbing the instructions, but it is assumed that candidates would have had prior practice in all these forms of question. It would also be desirable that the format should be used to cover a number of separate items. Whether or not it is desirable to include a sufficient number of such items in an actual test would have to be decided when the specification for the test was drawn up (see Chapter 7).

ITEM MATCHING

There is one other form of question suitable for objective tests which shares with the last form the disadvantage that its use may overweight one particular aspect of the syllabus. This is the matching form or item matching type.

Example 5.18

Consider the statements A–E below concerning an element x of a set $\mathscr{E}$ of which P, Q and R are subsets.

A $x \in P' \cap (Q \cup R)$
B $x \in (P' \cap Q) \cup R$
C $x \in P \cap (P' \cup Q)$
D $x \in P \cup (P \cap Q)$
E $x \in Q \cap (P' \cup R')$

In the following questions you are given a statement about x
D

and in each case you are required to give the key letter, A, B, C, D or E of the statement which can logically justify it.

(a) $x \in (P \cap R)' \cap Q$
(b) $x \notin P$
(c) $x \in P$
(d) $x \in P \cap Q$

Example 5.19
In each of the following sentences the blank can be filled by one of

A   median
B   mode
C   mean
D   standard deviation

For each sentence give the key letter A, B, C or D of the correct replacement in the sentence about a set of marks.

(a) The .... is unaffected by adding equal amounts to all the marks.
(b) The .... is always the mark gained by most candidates.
(c) The .... has always as many candidates with scores below it as above it.
(d) The .... can be calculated by summing the marks and dividing by the number of candidates.
(e) The .... is always less than the range of marks.

QUANTITATIVE COMPARISON

The last type of question which will be considered is one which is peculiar to mathematics, or at least to subjects with a mathematical content, and asks candidates to compare the magnitudes of two quantities. It may be referred to as a quantitative comparison type (Example 5.6 could be classified as this form). It could be worded as follows:

Example 5.20

In the following questions you are given two quantities (1) and (2). Answer

A   if $(1) > (2)$
B   if $(2) > (1)$
C   if $(1) = (2)$
D   if the information given is insufficient to determine which
    of A, B or C is true

(a) (1) 3^4 (2) 4^3
(b) (1) $\cos(\pi/4)$ (2) $\cos(-\pi/4)$
(c) (1) x^2 (2) y^2 where $x < y$
(d) In the distribution formed by removing the lowest 25%
    of a normal distribution
    (1) the median (2) the mean

6. *Item Construction for Objective Tests in Mathematics*

The purpose of this chapter is to show how an item writer might set about the task of constructing suitable items, and to describe some of the pitfalls which have to be avoided. It is assumed that he will have been told to construct items to cover a particular section of the syllabus and that he will know at what ability level his items should be set. This latter requirement imposes two restrictions on the item writer, not only must his items be constructed so that they are of the correct level of difficulty but they must also fulfil the given requirement of testing at a desired level of thought—knowledge, comprehension, application or the highest abilities.

A CONVENTIONAL QUESTION

Suppose it is required to write one or more items testing the pupil's mathematical ability in the particular context of the use of the Sine Rule. In a conventional examination this might well be done by means of a question such as Example 6.1.

Example 6.1
Three towns A, B and C are such that A is 18 km from B on a bearing of 035° and the bearing of C from B is 136°. If the bearing of A from C is 330°, how far is B from C?

The pupil's first task is to use the written information to make a mathematical model of the situation. This step involves of course the knowledge of the meaning of the term bearing and the ability to construct, at least roughly, angles of required size. Assuming the candidate has this knowledge and this ability he will draw a figure such as the one shown on next page.

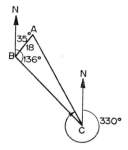

The pupil must next realize that he has sufficient information to 'solve' the triangle ABC and that the information he has is such that the Sine Rule is an appropriate mathematical tool to use. He must then recall the formula

$$\frac{a}{\sin A} = \frac{b}{\sin B} = \frac{c}{\sin C}$$

manipulate this formula to an appropriate form, calculate the sizes of the angles of the triangle and finally carry out the calculation in the process of which he has to show ability either in the use of logarithms or in the use of a slide rule. Analysed in this way, it is apparent that the question is attempting to do many things at one time. If the candidate fails in the first of these, the knowledge of the meaning of the word 'bearing', the conventional question can tell us nothing about his abilities in the other facets of the question. Objective testing gives us the opportunity to find out which parts of the question the pupil can answer.

It is obvious that it would be most undesirable to take as an objective item Example 6.1 with the addition of the 'correct' answer and four randomly chosen distracters. In fact the question as stated would form the basis of a number of objective items.

EQUIVALENT OBJECTIVE ITEMS

The problem that faces the item writer, then, is to construct a series of items to test some or all of the abilities implicit in Example 6.1. Suppose we look first at the testing of the term 'bearing'. Can a suitable item be devised to test this knowledge? Example 6.2 is a possibility.

Example 6.2

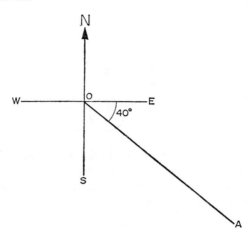

From the diagram the bearing of A from O is:
A 040°
B 050°
C 130°
D 230°
E none of these

The second stage of the problem—the ability to interpret the verbal question as a geometrical configuration—cannot be tested directly but it is possible to construct an item which will test an ability which on the surface would seem to be very close to this ability.

Example 6.3
Three towns A, B and C are such that the bearing of A from B is 035°, the bearing of C from B is 136° and the bearing of A from C is 330°. Which of the diagrams (a), (b), (c), (d) or (e), at the top of page 95, is most likely to show the positions of the three towns?

In the same way it is possible to write items which cover some of the other points tested in the original question. Some possible items are given in Examples 6.4–6.7.

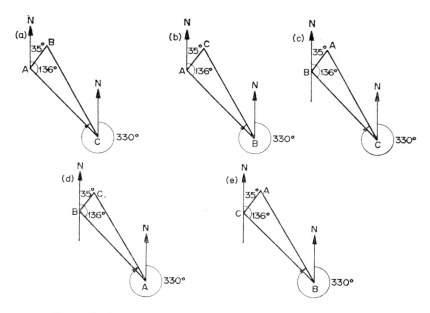

Example 6.4
In which of the triangles I–IV is sufficient information given to be able to calculate the length of XY?

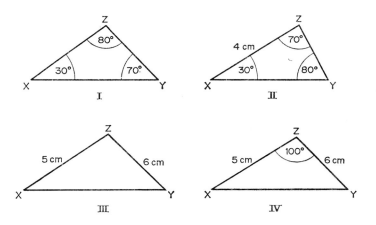

The usual multiple completion format would, of course, be used in this item.

Example 6.5

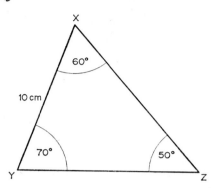

The length of YZ in cm is given by the expression

A   10 sin 50°/sin 60°
B   10 sin 60°/sin 50°
C   10 sin 60°/sin 70°
D   sin 60°/10 sin 50°
E   sin 50°/10 sin 60°

Example 6.6
The value of 10 sin 70°/sin 40° is

A   less than 10
B   between 10 and 13
C   between 13 and 16
D   between 16 and 20
E   greater than 20

Example 6.7
PQR is parallel to STV. The size of angle QWT is

A   50°
B   60°
C   70°
D   110°
E   not determinable from the given information

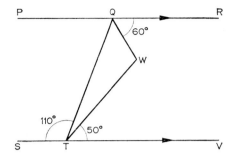

Figures

It will have become obvious from the items which have been given as examples both in this chapter and in previous chapters that it is very often desirable for a diagram to accompany an objective item. A diagram can often convey information in a form which is more comprehensible to an immature student than a written piece of information. The test constructor must, however, decide whether figures such as those of Examples 6.5 and 6.7 have to be drawn approximately to scale or not. If they are drawn to scale there is always the possibility of the candidate selecting the key by measurement of the figure rather than by the mathematical ability which it was intended to test. The same consideration arises in many items which are intended to test co-ordinate geometry and the balance of convenience would on the whole seem to be in favour of not attempting to make accurate drawings. This does not mean, of course, that it is right to use an inaccurate figure deliberately to mislead a candidate.

Accuracy

A second general problem which faces the test constructor is how far he should insist on mathematical accuracy. There will be numerous occasions on which to include all the mathematically necessary provisos in an item may result in a question so complex that it is incapable of interpretation by the type of candidate for whom it was intended. Is it justifiable in such cases to temper the wind of mathematics and allow a statement to appear which is not strictly accurate? Take the item in Example 6.8.

Example 6.8

The operator Δ is defined on the set of integers by

$$a \Delta b = \frac{ab}{a+b}$$

$(1 \Delta 2) \Delta 3$ is equal to

A $\frac{3}{7}$
B $\frac{6}{11}$
C $\frac{3}{4}$
D 1
E   none of these

This item should be rejected immediately since once the candidate has evaluated $1 \Delta 2$ as $\frac{2}{3}$ he is not entitled to go any further. Some of the candidates would therefore take as the correct answer E whereas others would carry on with the calculation and arrive at one or other of the given responses. However, even if the question is improved by substituting the set of real numbers for the set of integers, there is still the difficulty that $a \Delta b$ is not defined when $a + b = 0$. The question has therefore to be reworded as in Example 6.9.

Example 6.9

The operater Δ is defined on the set of real numbers by

$$a \Delta b = \frac{ab}{a+b}$$

for all a, b such that $a + b \neq 0$. $(1 \Delta 2) \Delta 3$ is equal to . . .

It is true that the added phrase is of no help to the vast majority of candidates but it probably has to be included for the sake of anyone whose attitude might be 'This question is mathematically meaningless. I cannot answer it.'

Simplicity

Provided it can be done without a loss of accuracy, the wording and the type of calculation called for should be kept as simple as possible. An objective test by its nature must contain much more for the candidate to read than a conventional paper, and it is necessary to prune any unnecessary words ruthlessly. Thus in Example 6.7 it would be usual in a conventional paper to specify the sizes of the given angles in words as well as in the diagram. In an objective item the words can well be omitted. Consider the following example.

Example 6.10
y varies inversely as x^2. If $y = \frac{1}{2}$ when $x = 16$, the value of y when $x = 32$ is

A $\frac{1}{8}$
B $\frac{1}{4}$
C 1
D 2
E none of these

If it is intended to test variation then this item is no better than the same item with the x values replaced by 2 and 4 and the same responses. If the arithmetic has to be complicated in order that the item should not be too easy, then the item writer must realize that what is being tested is not variation but arithmetic. There may be occasions on which we want to test arithmetical manipulation and there can be no objection to doing so, but it is surely better to test this separately from the test of mathematics.

The stem

The stem is that part of the item which poses the problem. It may be either in the form of a question or in the form of an incomplete statement; there is no particular reason for preferring one rather than the other. Thus the same item could appear as either one of the next two examples.

Example 6.11
19 257 $= 3 \times 7^2 \times 131$. In what base can 19 257 be written as 110 100?

A 5
B 6
C 7
D 8
E 9

Example 6.12
$19\,257 = 3 \times 7^2 \times 131$. $19\,257$ can be written as $110\,100$ in base:

A   5 etc.

In mathematics it is desirable that so far as possible the stem of item should pose the problem so that the candidate, having read the stem, is aware of the nature of the problem. If possible the items should be such that the good candidate, having read the stem, has only to select from the responses what he already knows is the key.

Items such as Example 6.13 have no place in a properly constructed objective test.

Example 6.13
Which of the following is false?

A   The diagonals of a parallelogram are equal
B $(x + 3)^2 = x^2 + 6x + 9$
C $\sin 30° = \frac{1}{2}$
D $15_{\text{six}} + 23_{\text{six}} = 42_{\text{six}}$
E   The line joining $(0, 3)$ and $(1, 6)$ has equation
$$y = 3x + 3$$

The stem should also be concerned with one concept only, not, as in the last example, with a number of different pieces of mathematics. This concept need not be a simple one, but it is good practice after writing an item, to ask what it is that the item is trying to test. If the item tests a number of unrelated ideas then it should be discarded.

It is possible for the stem to be framed in a negative form, but if so the negative should be emphasized by underlining or by a different form of print.

Example 6.14
If the magnitude of a vector $\underline{u}$ is 3 units and the magnitude of a vector $\underline{v}$ is 4 units, which of the following can NOT be the magnitude of $\underline{u} + \underline{v}$?

A   1
B   3

C 5
D 7
E 9

Example 6.15

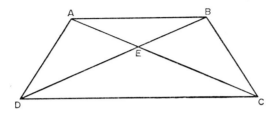

ABCD is a quadrilateral whose diagonals intersect at E. Which one of the following is NOT a sufficient reason for the statement 'AB is parallel to DC'?

A   angle ABD = 60° and angle BDC = 60°
B   AE : EC = BE : ED
C   angle ABC + angle BCD = 180°
D   AB : BE = DC : DE
E   angle ADC = angle BCD and AD = CB.

While items such as these may on some occasions be the only way of asking a particular question and should not cause the candidate undue difficulty because of the wording, items which include a double negative, either both in the stem or one in the stem and one in the responses should be avoided; in Example 6.15 it would not be reasonable to frame the stem as follows: 'Which one of the following is NOT a sufficient reason for the statement "AB is NOT parallel to DC"?' In the same way it would be undesirable to ask the question in Example 6.16.

Example 6.16
If $x, y \in R$ and $x^2 + y^2 \neq 0$ then
1. $x + y \neq 0$ 2. $x^3 + y^3 \neq 0$ 3. $x^4 + y^4 \neq 0$

(the usual multiple completion format being used).

There is at least the possibility that this is as much a test of interpretation as a test of mathematics.

Responses

The distractors, that is the incorrect answers to the items, must be chosen so that they have a positive appeal to candidates who lack the mathematical knowledge or skill which it is intended to test. The simplest way to find suitable distractors is to give the item to a class of the appropriate age and ability group in a completion form and note the incorrect answers which arise.

Since, as has been stated, the calculations involved should in most cases be kept as simple as possible, these incorrect answers are most likely to arise from misinterpretation of the question. Thus in Example 6.10, the distractors are obtained from the calculations arising from y varies as x, y varies inversely as x and y varies as x^2. The responses 'none of these' is worth including in items involving arithmetical answers since it allows for candidates' errors in computation. It is, however, essential that 'none of these' should sometimes be the correct answer if candidates are not to be accustomed to rejecting it without thought. It would not be an appropriate fifth response to the item in the following example, since it would penalize the candidate who correctly arrives at the result $\frac{1}{2}(P/\pi r - r)$.

Example 6.17
If $P = \pi r(r + 2s)$ then s equals

A $\dfrac{P - \pi r^2}{2r}$

B $\dfrac{P}{\pi r^2} - 2\pi r$

C $\dfrac{P}{2\pi rs}$

D $\dfrac{P - r}{2}$

Item writers should try to avoid giving inadvertent clues in an item, for example by making the desired response much longer or more explicit than the distractors.

Example 6.18
The value of $\displaystyle\int_{-1}^{1} 1/x^2 \, dx$ is

A -2
B 0
C 2
D 4
E none of these since $1/x^2$ is undefined when $x = 0$

When the stem is in the form of an incomplete statement, the responses should all follow grammatically from the stem. Otherwise the candidate may be given unintentional clues to the key, as in Example 6.19.

Example 6.19
In the expression 2^n, the 'n' is called an

A index
B coefficient
C base
D power
E root

This is a very obvious example; Example 6.20 is perhaps not so obvious but has the same kind of fault.

Example 6.20

The points $(1, 2)$, $(2, 5)$, $(3, 4)$ and $(2, 1)$ are the vertices of a

A square
B rectangle
C rhombus
D parallelogram
E kite

With no knowledge of co-ordinates the candidate can rule out responses A, B and C since each of these would also imply D. There is therefore only a two-choice option. Admittedly the candidate is using mathematics, but not the mathematics which the item writer intended.

The responses should be arranged in some systematic way. If the responses are all numerical they might be in ascending or descending order of magnitude.

Ability Level

It was stated at the beginning of this chapter that the item writer will be given instructions to write items testing all levels of ability. This can be a very difficult task since even conventional examinations have tended to concentrate on the lowest levels of ability. Not only are high-level items difficult to devise but it is difficult to be sure at what level an item is testing. Whether an item tests knowledge of the highest ability depends not only on what the candidate has been taught but also on how he has been taught. The same item can test different candidates on different levels.

Example 6.21
With how many zeros does the number 27! end?

A   4
B   5
C   6
D   7
E   none of these

This example can operate at three different levels depending on the ability and the previous experience of the pupil who answers it. At the lowest level it can be answered by routine multiplication of 27!, at the highest level by the appreciation that the problem is equivalent to being asked the number of times 5 can be divided into 27! At an intermediate level comes the performance of a candidate who has previously been asked a somewhat similar question. As an illustration of the different levels at which questions may be written consider the items in the following four examples.

Example 6.22
The statement $\log_r s = t$ is equivalent to

A $r = s^t$
B $r = t^s$
C $s = r^t$
D $s = t^r$
E $t = r^s$

This item is a pure knowledge question, testing whether the candidate knows the definition of a logarithm in terms of indices.

Example 6.23
The value of $\log_2 64$ is

A $\frac{1}{6}$
B 5
C 6
D 8
E 32

This item might correctly be put in the second category—comprehension. The candidate has first to realize that he has to recall the definition of a logarithm, secondly to apply this definition to obtain the equation $2^x = 64$ and finally to see that the solution of this equation is $x = 6$.

Example 6.24
2^{100} is

A less than 10^{20}
B between 10^{20} and 10^{30}
C between 10^{30} and 10^{40}
D between 10^{40} and 10^{50}
E greater than 10^{50}

This would be classified as an application item, assuming that the candidate has not had previous experience of this type of question. The candidate has to realize that this item is one which should be dealt with by the use of logarithms; he has to know that the greater of two numbers has the greater logarithm; he has to know that $\log 2^{100}$ equals $100 \log 2$; use tables to evaluate this as approximately 30.1; realize that this is the logarithm of a number between 10^{30} and 10^{40}.

Example 6.25
Consider the solution of the equation
$$\log_x (2x^3 + x^2 - 2x) = 4.$$
$$\log_x(2x^3 + x^2 - 2x) = 4 \Rightarrow 2x^3 + x^2 - 2x = x^4$$
$$\Rightarrow x^4 - 2x^3 - x^2 + 2x = 0$$

The equation $x^4 - 2x^3 - x^2 + 2x = 0$ has solutions $x = 0$, $x = 1$, $x = -1$ and $x = 2$.

The equation $\log_x (2x^3 + x^2 - 2x) = 4$ has solutions

A $x = 0, x = -1, x = 1$ and $x = 2$

B $x = 0, x = 1$ and $x = 2$ only

C $x = 1$ and $x = 2$ only

D $x = 2$ only

E   some other values of x

Here the candidate is being asked to evaluate a proof and the item would be placed in the category of higher abilities. He has to realize that the implication in the first line can only be reversed for values of x for which $\log_x (2x^3 + x^2 - 2x)$ is a meaningful expression, i.e. provided x is greater than 0 and not equal to one.

SOURCES OF ITEMS

It is not easy to write good objective items at a level above knowledge. Even knowledge items require care and attention to the few basic rules that have been given. Few item writers will ever acquire the ability to produce items immediately to order: ideas have to be garnered from a variety of sources as they are met, and kept for subsequent use. This difficulty in producing original ideas for high-level items is not, however, a difficulty that does not already exist in traditional forms of examination. The new feature is that where before a setter might only have to produce one or two such questions, now the test constructor will require ten or twenty. This need emphasizes the desirability of co-operative work among a number of teachers in producing a test. It means also that good, high-level items, when they have been found, deserve more use than one examination.

Most ideas will come to the item writer in the course of his normal teaching. The pupil who asks a question in a different way from the others may provide the start of a train of ideas which will produce an item; a wrong piece of reasoning in an examination may contain the germ of an idea; textbooks and books of mathematical puzzles can suggest situations from which suitable items can arise.

7. *Compiling an Objective Test*

The first step in constructing an objective test is to draw up a specification for the test. This specification should cover the three factors:

(i) the weighting to be given to the different parts of the syllabus,
(ii) the weighting to be given to the various objectives of mathematics teaching,
(iii) the number and type of question to be used.

FORM

It is unlikely that any test will contain all the types of question considered in Chapter 5—in particular it is generally wise to refrain from using true–false items—and in small-scale examinations there is a good case for using more completion items than would be found in a national examination. Pupils require some training in examination techniques. In dealing with an unsophisticated group—unsophisticated, that is, in the sense of being unused to objective tests—it is · reasonable to restrict the test to two or at most three types of item. As examinees acquire greater sophistication it will be possible to introduce additional types, but a reasonable test can be constructed using only multiple choice, multiple completion and situation set items.

The length of the test depends on a number of factors; for example, whether the test is to be complete in itself or to form part of a larger assessment, and the age and ability of the group which is being tested. However, if a test is to be worthwhile it should contain at least forty or fifty items and it is probably realistic to think in terms of an hour or an hour and a half for such a test. It should be realized that with this type of test the candidate is having to spend his time in mathematical thought, and not in manipulative exercises as has sometimes been the case in the traditional form of examination. It follows that intensive thought is required of the examinee and it is not realistic to think of a fifteen-year-old being required to spend two to two and a half hours at this task. This is not to deny that there is a place in the classroom for short objective tests, say fifteen items to be done in twenty minutes which can be used at the end of each section of the scheme of work.

SPECIFICATION GRID

For a test designed for a particular purpose with a particular ability group, it is convenient to draw up a specification grid for the test. This, as was mentioned in Chapter 3, is a two-way table of abilities and syllabus. It is obviously undesirable that all the items on, say, algebra should test only knowledge and all the items on geometry should test comprehension. It is better to have a spread of abilities among the syllabus sections. A specification grid for a fifty-item test of mathematics for seventeen-year-olds might have the following pattern, where the abilities A, B, C and D are as specified in Chapter 4.

Abilities Syllabus	A	B	C	D	Total
Geometry	3	6	4	2	15
Algebra	5	4	4	2	15
Trigonometry	2	2	2	1	7
Calculus	4	4	3	2	13
Total	14	16	13	7	50

The contents of the grid will of course vary from different age groups and ability levels and for tests designed for different purposes. Thus a diagnostic test might well assign greater weight to the lower abilities, a prognostic test greater weight to the higher abilities. Whatever weights are assigned to the various cells of the grid, it is essential that the test compiler should have some specification to work from.

SYLLABUS

Having constructed or having been given the test specification, the test constructor must then decide how the items in, say, algebra are to be distributed over the syllabus. Most examinees will be working to a syllabus laid down by an external authority, and the test constructor has to take the syllabus and decide what weighting should be given to each of the topics. At this stage it is not possible to be precise about the actual weighting since this will depend on the difficulty and inter-correlation of the items, but it would not be too imprecise to take the weighting of the sections as proportional to the number of items used.

The test constructor must make a syllabus breakdown into as small sections as possible. Once this has been done he may decide that some of the sections are not well suited to objective testing—it would be difficult to test, for example, a candidate's ability to draw the circumcircle of a triangle as distinct from his knowledge of how to draw it. At this stage there will probably be well over fifty different syllabus sections which it is desired to test, but it is unlikely that each of these sections will be of equal importance. Few mathematics teachers would equate, for example, the periodicity of trigonometrical functions with the value of sin 45°. It becomes obvious, therefore, that even assuming a breakdown into fifty syllabus sections, the desire to give differential weighting to different sections would require possibly 150–200 items. Such a test would be impossibly long and the constructor has to select from the possible field of syllabus sections. This selection is a subjective choice which will vary with different examiners.

ABILITY LEVELS

As well as giving a spread over different parts of the syllabus an examination should, as has already been stated, give a spread over the different ability levels identified earlier. Again, it must be emphasized that the assignment of items to different ability levels is a subjective judgment on the part of the test constructor or item writer. Even if there is a clear distinction between knowledge and comprehension—and it is more likely that the abilities form a continuum rather than a set of discrete classes—it is true that an item can only be classified for a group of candidates who are known to have had an identical training. Consider the item in Example 7.1.

Example 7.1
For how many integers n is $12/(n^2 - 5n + 6)$ an integer?

A 0
B 1

C   2
D   3
E   more than 3

This would, if the candidate has not seen such an item before, be classified as higher abilities. If he had been given previous training on items such as this, it would be a knowledge item; if he had seen such items before but only using expressions such as $12/(n-3)$, it would be a comprehension item. The test constructor can only use his experience to arrive at a classification which he believes will be valid for the majority of candidates. Where the test is being constructed for internal use in a class or school, this classification should be more reliable than it can be on a larger scale since the teacher has more knowledge of the background of the candidates. Again, however, this is a subjective judgment. Nevertheless, although this classification cannot be precise, there is great value in attempting to make it.

ITEM-VETTING

After items have been written and before they are used either in a test or in a pre-test, it is almost essential that they should be subject to review by one or more independent persons. This process of review is called vetting or shredding the item.

An item vetter has to consider every item in the light of all the points which were covered in the chapter on writing items. He should not be given the key in advance and will find it helpful to keep a check-list which can be used to review each item. A possible list for such a purpose might be:

1. What is the key?
2. What syllabus section does the item test?
3. What ability does the item test?
4. Is the mathematics sound?
5. Are there clues to the key?
    (a) length of key
    (b) specific determiners
6. Are the distractors plausible?
7. Can the wording be improved?
    (a) factoring out

(b) over-sophistication

(c) elimination of unnecessary detail

8. Is it reasonable to expect an average candidate to answer the item in the time allowed?

9. Where appropriate are the responses in systematic sequence?

10. Is there more than one defensible answer?

SELECTION OF ITEMS

If the final version of a test is being compiled, rather than a pre-test, the items will already have been pre-tested and statistics such as those considered in the next chapter will be available. From the pre-tested items, the test constructor then selects the required number, say fifty, to satisfy the requirements of the grid and to meet the desired criteria with respect to facility values and discrimination values. For an achievement test he will be looking for a mean facility value probably in the region of 0.6, with a spread between, say, 0.35 and 0.85. Items outside this range are likely either to be so easy that they will play little part in distinguishing between candidates or so difficult that they will discriminate only at the extremes of the population being tested. Probably the spread of facility values should not be uniform but have, say, thirty items in the range 0.5 to 0.7, with the remaining twenty evenly divided between the more difficult and the easier. Within this framework the selection of items will be a subjective decision by the test constructor. For a pre-test this selection can only be the result of the use of the compiler's professional skill.

DESIGNING THE PAPER

Once the items have been selected it is necessary to assemble them into a test paper which will be presented to the candidates. A preliminary decision which must be made is whether the candidates should use a separate answer sheet or show the answers on the question paper. The advantage of the latter method is that there is no danger of the candidate putting the answer to, say, item 20 in the answer space for item 21. Against this must be put the disadvantages that marking becomes much more laborious—as does any analysis of the answers—and that there is then no possibility of re-use of question papers. It is probably therefore worthwhile accepting the small risk of wrong placing of answers for the advantages to be gained from a separate answer sheet.

THE ANSWER SHEET

The answer sheet can either require the candidate to write the letter of the key or to mark a space corresponding to the key. The latter method is slightly to be preferred because of the facility of correction, and has the additional advantage that the candidate will meet this system in any external examination where the answer sheet is designed to be read directly by a machine.

A suitable answer sheet can be made up on A5 paper (210 mm × 148 mm)—to provide space for fifty five-option fixed-response items. A possible format for such a sheet might be as follows.

Name_____ Mark_____

Class_____

Put a tick (✓) in the box corresponding to the correct answer.
To delete an answer cross out the tick (✗). If two or more
ticks are left for any item, no mark will be awarded for that
item. Your score will be the number of items you answer correctly.

1. ☐ ☐ ☐ ☐ ☐   26. ☐ ☐ ☐ ☐ ☐

2. ☐ ☐ ☐ ☐ ☐   27. ☐ ☐ ☐ ☐ ☐

THE QUESTION PAPER

The fifty items have to be arranged in the question paper in some order. The compiler has the choice of using a subject order, that is, put all the algebra together then the geometry etc., or in order of difficulty or of compromising between the two. Putting the items in order of difficulty has the advantage that it is less likely that the candidates who are of median standard of ability will spend too long on an early item and thus leave himself short of time for the later, perhaps simpler items. It has the disadvantage that the candidate has to re-orientate himself at each item from one branch of mathematics to another. There is therefore much to be said for a compromise in which the items are first sorted into three categories—easy (facility values greater than 0.7), moderate (facility values between 0.5 and 0.7) and difficult (facility values less than 0.5). Within these categories items can be arranged in subject order with some latitude allowed in order to make best use of the available space. In passing it may be mentioned that the bulk of an objective test will be much greater than that of the more traditional test. A fifty-item test, making adequate use of diagrams, cannot be expected to occupy less than twelve pages and can quite easily run to sixteen or seventeen. This consideration is a further reason for the re-use of tests.

CHECKING THE PAPER

When the paper has been compiled it is essential that it should be checked, preferably by someone who has not seen the items before. This checker should work through the paper, taking special care to note any possible ambiguities in the items and any cases where the stem of one item may contain information which will be of assistance to candidates in answering another item. This will be the first time that the paper has been seen as a complete entity, and it is not until this stage that any such interrelationships will become apparent. Finally it should be stressed that the inconvenience to the marker of a wrong answer will be much greater in a fixed-response test than in a conventional examination.

SUMMARY

Although these suggestions for the actual construction of the test paper have been few, it cannot be over-emphasized that the proper construction of a test is far more difficult than the writing of individual items. The constructor will continuously be involved in value judgments, although it should be the aim of the person or group of persons who lay down the specification for the test to make the necessity for such value judgments as small as possible. The validity of the value judgments can only be assessed after the test has been administered. Some of the judgments will then immediately be liable to assessment but others can only be assessed much later. If, for example, a test is designed to assess candidates' ability to satisfactorily complete a university course, the judgments implicit in the test construction can only be assessed after the completion of such a course.

8. *Item Analysis*

The introduction of objective testing allows a more scientific approach to the construction of examinations. It is possible to design a test to have a desired characteristic in respect of the distribution of marks that can be expected from its use. To achieve this it is necessary to have a preliminary trial—or pre-test—of the material which it is intended to use. As well as allowing the proper design of the final test, such a trial allows undesirable items to be eliminated before the final test is assembled.

PRE-TEST POPULATION

The population for the pre-test must be selected so that it has the same characteristics as the population for whom the test is intended. If an examination is required for, say, the top 30 per cent of the 16-year-old age group, there is little point in testing it out on a group of 17-year-olds who have continued the study of mathematics. Not only will the age difference make different attitudes to the items likely, but also it is probable that the 17-year-olds will form a selected group—if only self-selected—which would make interpretation of the statistics obtained very difficult. For a national examination, the pre-test group should if possible mirror the population for the actual examination in age, sex, type of school and course of study. This requirement, coupled with the necessary requirements of security, involves the examining authority in some difficulty.

The size of the pre-test population need not be large. Provided it has been selected without bias, a sample of 250–400 should provide sufficiently reliable statistics to allow the construction of a satisfactory examination, and even smaller samples can provide worthwhile information. When the problem is translated to the level of the individual teacher attempting to construct an objective test, there are additional difficulties. It is unlikely that the pupils in any one school would form a suitable pre-test population if it is desired to construct a test which would be comparable with one constructed for use on a large scale. Items which would give acceptable statistical characteristics at national level might well appear too easy or too difficult or to lack discrimination when given to pupils in a particular school. It must also be realized that it is unlikely that more than 50 per cent of the original items will be acceptable so that in order to produce a test of 40 items—giving adequate syllabus coverage—it will probably be necessary to provide over 100 items for testing. Unless it is absolutely unavoidable a teacher is probably well advised not to attempt this task on his own but to try to organize a pool of colleagues on an area basis to collaborate in the effort involved.

FREQUENCY OF RESPONSES

The pre-test of items is intended to produce three pieces of information about the items. The first of these is the number of candidates who choose each of the responses to the item, and the number who omit the item. Some would advocate that in addition it is desirable to note the number who do not reach the item in the time allowed, but it has been assumed throughout this book that the length of the test will be such that all or at least the great majority of the candidates will have sufficient time to complete the test, and it is therefore not worthwhile to distinguish between those who omit and those who do not reach an item. The number of candidates who do not respond to an item will depend to some extent on the policy adopted on correction factors.

A five-response item might well produce a frequency of responses when administered to a population of 400 of

A 220*   B 35   C 47   D 42   E 46   O 10

(The asterisk denotes the key and O gives the number of candidates omitting the item.)

In this item each of the responses has been satisfactory in that it has attracted a reasonable percentage of candidates. On the other hand a pattern such as

A 220*   B 139   C 7   D 9   E 15   O 10

would be unsatisfactory in that the item has functioned for 90 per cent of the candidates as a two-choice item, with all the inherent disadvantages of such items. The criterion for rejection on this aspect of the pre-test result is a value judgment—as are all the criteria which will be given in this chapter—but it is suggested that a five-response item is not suitable for use without amendment unless each of the four distractors attracts at least 3 per cent of the candidates. Alternatively it might be sufficient to stipulate that three of the four distractors should each attract at least 5 per cent of the candidates. With a four-response item the criterion might be that at least 5 per cent of the candidates should choose each distractor; the alternative would not be appropriate in this case.

CORRECTION FACTORS

The values of the other two statistics which will be considered in this chapter depend upon the policy which is adopted in finding the candidate's total score, and it is therefore appropriate to consider this policy before going on to look at the statistics. Any fixed-response test is liable to errors in assessment due to guessing, and it is necessary to decide whether the marking system should be adjusted so as to attempt to compensate for this error. If a 100-item test were given to a large number of candidates who answered the items randomly, and one mark were awarded for each correct answer the marks would theoretically approximate to a normal distribution with mean $100/n$ where n is the number of responses to each item. The standard deviation of the set of marks would be $10\sqrt{(n-1)}/n$. The significance of this can be shown in the following table which lists for $n = 2, 3, 4$ and 5 the percentage, to one decimal place, of the candidates who would achieve scores of at least 20, 30, 40, 50, etc.

n	Mark								
	20	30	40	50	60	70	80	90	100
2	100	100	97.7	50	2.3	0	0	0	0
3	99.8	76.1	7.9	0	0	0	0	0	0
4	87.6	12.4	0	0	0	0	0	0	0
5	50	0.6	0	0	0	0	0	0	0

In the case of a 50-item test marked on the same basis the corresponding table would be:

n	Mark								
	10	15	20	25	30	35	40	45	50
2	100	99.8	92.1	50	7.9	0.2	0	0	0
3	97.7	69.1	15.9	0.6	0	0	0	0	0
4	79.3	20.7	0.7	0	0	0	0	0	0
5	50	3.8	0	0	0	0	0	0	0

It will be seen from these tables that provided a sufficient number of items are given, and provided each item has four or five responses, the probability of a high score by guessing is not large. In practice an objective test would form only part of an examination, thus reducing the effect of guessing. However, it also shows that a pass–fail examination should be designed so that the pass mark is well above the mark which is likely to be gained by guessing—such an examination with a 'pass mark' of, say, 30 per cent would be completely unsatisfactory.

Where a correction formula is desired, the simplest form to use is to give a mark of $+1$ for a correct answer, 0 for an unanswered item and $-1/(n-1)$ for a wrong answer, where n is the number of responses to each item. The theory of this correction factor is that every wrong answer is the result of random guessing. In a five-response test, if a candidate guesses then for every four wrong responses he will on average make one correct response. Therefore in a 50-item test, if he has 34 correct responses and 16 wrong responses the assumption is made that 4 of the correct responses were the result of guessing. This method of scoring ignores the fact that in a well-constructed objective test, wrong answers are probably not the result of random guessing. Distractors are selected because of their positive appeal to those candidates who lack either the ability or the knowledge that is being tested. There is therefore no reason to suppose that a candidate who has sixteen wrong answers has guessed the answers to a further four items correctly. He may not have guessed the answers to any items. Use of the correction formula would therefore tend to over-correct the candidate's score. On the other hand, candidates may still guess without guessing completely at random. To the extent that they can eliminate certain of the distractors they will be choosing from fewer than the intended number of options. An appropriate correction formula to arrive at the number of items to which the candidate *knew* the answer might therefore be $1/(n-2)$ or $1/(n-3)$. To this extent the conventional correction formula under-corrects.

Assuming all the candidates answer all the questions there is no effect on the rank order of candidates by using or not using a correction formula since the candidates' total score is $C - W/(n-1)$ where C is the number of correct answers and W the number of wrong answers. If $W = N - C$ where N is the total number of items the total score reduces to $(nC - N)/(n-1)$. Thus the total score is a linear function of the number of correct answers.

It would seem therefore that in a test which is of such a length that all or nearly all of the candidates can complete it the decision to use or not to use correction factors will depend not on any hope of increasing the accuracy of the score but on the attitude which the test-compiler wishes to induce in the testee. The use of a correction factor will probably discourage the unsophisticated candidate from guessing. The apparent undesirability of guessing is due to the loaded nature of the word 'guessing'. Guesses, in this context, are very seldom random. They tend to be a combination of half-remembered facts and intuitive reasoning. With most 15–17-year-olds the problem is not to discourage guessing but to encourage the giving of an answer when the candidate is less than 100 per cent sure of its correctness. Each test user will have to decide his policy for himself, but it will be assumed in the rest of this chapter that no corrections have been made to the raw scores. The criteria given must be interpreted in this situation.

FACILITY VALUE

The facility value of an item is that fraction of the candidates who give the required response. Thus if 60 out of 240 candidates give the desired response the facility value is 0.25 (some writers prefer to use percentage and would quote this as 25 per cent). It should be noted that this fraction may or may not be the fraction who *know* the correct answer. If an item gave the pattern of responses

A 60*   B 40   C 35   D 45   E 50   O 10

almost certainly less than $\frac{1}{4}$ of the candidates knew that A was the required response. We cannot tell what this fraction should be.

Some writers advocate using $(60 - 170/4)/240 = 0.07$, others advocate $(60 - 50)/240$. The first of these is obtained by the correction formula outlined above: the second by the argument that candidates who answer can be divided into five sets, those who know the correct response is A, those who know it is either A or E, those who know it is A or E or D, those who know it is A or E or D or B, and the remainder. It is easily shown on this argument that the number who *know* that A is the required response is the number choosing A less the number choosing the next most popular response.

It is reasonably clear that we would not want to select—if the selection were done on the grounds of facility values only—items with as low a facility value as 0.25. This decision cannot be made without reference to the purpose of the examination. Where the test is used to assess mastery of the subject matter, facility values of the order of 1 are clearly desirable. Similar values would also be desirable where the test was used as a diagnostic test. It is usually stated that for maximum spread of results a facility value of 0.5 is required for each item. This could, however, lead to a situation where the candidates were simply divided into two equal groups, one with a mark of zero, the other with the maximum possible mark. Obviously this would be an unreal situation as an assessment, of mathematical ability. Possibly what should be aimed for is a spread of facility values from a lower limit of about 0.25 to an upper limit of about 0.85. Items with facility value above this are of little use as part of a measuring instrument. There is a case for including such items at the beginning of a test to allow candidates to make a confident start. Where the test is being used in a school situation it is necessary to guard against the danger of discouraging pupils by giving items which only one half of them can get correct, and in this situation it is reasonable to accept the lack of discrimination consequent on the inclusion of items which have higher facility values. If the test is to be used as the only constituent of a pass–fail examination and it is intended to pass, say, 75 per cent of the candidates, then the test should be designed to give an average facility value of 0.5 for those candidates who will be at the foot of the top 75 per cent. This would imply an overall facility value of somewhere about 0.6.

DISCRIMINATION INDEX

A suitable facility value is not the only criterion that objective items must satisfy. An item might have a facility value of 0.5 and still be unacceptable if the candidates who gave the correct response were not the candidates with most ability in the subject being tested. We require some measurement of the ability of an item to pick out the 'good' candidates. This in turn means that we require a measurement of the candidate's ability in mathematics or more precisely in that aspect or aspects of mathematics which the item is designed to test. This measurement or criterion cannot be precise since the existence of an exact criterion would imply the existence of a perfect device, and if such existed it would be used instead of the objective test. The criterion then can only be an approximation and various approximations can be used. The most readily available is the candidate's score in the complete test (or possibly in the complete test except for the item which is being assessed). Use of this has the advantage that pre-testing becomes a self-contained process in which the criterion can be generated in the course of marking the test. It has the disadvantage that it must be assumed that the test as a whole is a satisfactory instrument for assessing ability. This can be justified by the argument that while the test assemblers may make errors on the suitability of individual items, their professional judgment is unlikely to be at fault on the test as a whole. It is not unreasonable to accept this since all examinations in the past have been based on the acceptance of this judgment. This internal criterion can be refined by dividing the test into sub-tests each intended to assess one facet of mathematical ability and thus obtaining different criteria for items intended to assess knowledge, techniques, comprehension, application and higher abilities. Similarly different criteria could be established for geometry, algebra, trigonometry, calculus, etc. However, a reliable criterion requires generally the use of a reasonable number of items, say thirty, and the scope for establishing such varying criteria is therefore limited by the necessity of keeping the test within acceptable limits of length.

An alternative method is to use an external criterion. This might be performance in a traditional form of examination set at the same time as the objective test. This, of course, assumes that both the traditional and objective tests measure the same abilities, and this is perhaps open to doubt. (This doubt is one reason for not advocating purely objective examinations as a means of assessment.) Once a satisfactory objective test has been constructed, it can be used to provide a criterion for pre-testing further items provided that there has been no change in the syllabus or in the the relative weighting of the syllabus sections and ability levels which it is desired to assess.

When the criterion has been decided there are various ways in which it can be used to arrive at an index of discrimination. Which one is used will depend on the facilities available and, as might be expected, the more refined the discrimination index, the more work is involved in calculating it. The simplest method is to divide the test population into three numerically equal groups by their criterion scores. The facility value for an item can be calculated for each of the three groups. If these are F_t, F_m and F_b a possible discrimination index would be $F_t - F_b$. If this index is used the acceptable lower limit would be around 0.3. Alternatively two indices could be calculated, $F_t - F_m$ and $F_m - F_b$. In this case it would be reasonable to accept items where each of these exceeds 0.1. Still working with these three groups, an alternative method has been advocated. This requires the calculation of two phi-coefficients; one for the top third against the rest and the other for the bottom third against the rest. The average of these two coefficients can be used as an index of discrimination.

A refinement on this three-way split is to divide the population into five groups. The numbers in each group giving each response can then be tabulated thus:

Response	Group					Total
	I	II	III	IV	V	
A	70	60	50	40	30	250
B	5	10	10	15	10	50
C	4	7	10	10	15	46
D	1	3	5	10	10	29
E	0	0	5	4	12	21
O	0	0	0	1	3	4
Total	80	80	80	80	80	400

E

Such an array allows any non-functioning response to be identified, and would also show any tendency for a response other than the desired one to be selected by the better candidates.

In each of these methods each category of candidate is looked at in turn to find which response they select. If, instead, each response is looked at in turn to find what kind of candidate selects it, a slightly different method of displaying the results is obtained. This consists of showing for each response the average criterion score of the candidates who selected that response. Normally the key would be chosen by candidates whose average score was higher than that of the groups choosing any of the distractors.

In each of the last two methods it is convenient to have one statistic to express the discrimination of the item. For this purpose the biserial correlation or the point biserial correlation can be used. The former has the advantage that it is not so affected by differences in facilities between two items. If a biserial correlation is used, items with a value greater than 0.3 might well be considered suitable for use in a test. There are various short-cut methods available for calculating both facility values and indices of discrimination. Davis, for example, has produced a composite table which allows the facility value and a discrimination index which is closely connected with the biserial correlation, to be read off from a knowledge of the fraction of the top and bottom 27 per cent of the population who give the required response.

SPECIMEN ITEM ANALYSIS

All the statistics in this section refer to a pre-test conducted on a group of 150 sixteen-year-olds in the top 30 per cent of the population. The criterion score is the score on the test as a whole.

Example 8.1

Which of the points, A, B, C, D or E is the image of P under the dilation $(O, -\frac{1}{2})$?

Response	Frequency	Criterion score		
A	3	17.33		
B	46	25.61	F.V.	0.55
C*	83	28.68	r_{bis}	0.45
D	9	21.00		
E	3	24.33		
O	6	17.83		

This item has a suitable facility value and discrimination index. The correct answer is being chosen by, on the whole, the better candidates. However, two of the distractors are not functioning. The item would probably be rewritten with suitable replacements for A and E. Possibly a distractor might be a point at 1.5 to trap the candidate who confuses dilations and translations, a second possibility would be the point at 3.

Example 8.2
The number of integers x for which $7/(2x - 1)$ is an integer is

A   1
B   2
C   3
D   4
E   greater than 4

Response	Frequency	Criterion score		
A	32	29.6		
B	26	29.0		
C	22	24.3	F.V.	0.16
D*	24	27.3	r_{bis}	0.19
E	35	23.2		
O	8	21.1		

This item has proved much too difficult for the group being tested. Moreover the correct answer is not chosen by the best candidates. There is no option but to reject this item completely.

Example 8.3
$P = \{2, 3, 5, 7\}$, $Q = \{1, 3, 5, 7, 9\}$; $\{R = 2, 4, 6, 8\}$ are subsets of $\mathscr{E} = \{1, 2, 3, 4, 6, 7, 8, 9\}$. $P \cap Q \cap R'$ equals

A $\{1, 2, 3, 5, 7, 9\}$
B $\{3, 5, 7\}$
C $\{2\}$
D ø
E none of these

Response	Top	Middle	Foot	Total	
A	1	2	4	7	
B★	37	43	17	97	
C	4	5	4	13	F.V. 0.65
D	0	7	7	14	Discrimination
E	1	8	8	17	Index 0.42
O	1	1	0	2	
Total	44	66	40	150	

The analysis this time has been done by taking the group in three sections, top 27 per cent, foot 27 per cent and the remainder. It will be seen that each of the distractors is functioning and that the facility and discrimination are acceptable. Statistically this is a good item.

Example 8.4
The gradient of the line $3x + 4y + 7 = 0$ is

A $-\frac{7}{3}$
B $-\frac{4}{3}$
C $-\frac{3}{4}$
D $\frac{3}{4}$
E $\frac{4}{3}$

Response	Group					Total
	1	2	3	4	5	
A	0	0	16	9	7	6
B	0	3	13	17	29	13
C	4	14	10	0	25	10
D*	85	76	52	37	18	53
E	11	7	6	11	7	9
O	0	0	3	26	14	9

For this analysis the complete group was split into five groups, as nearly as possible numerically equal, on the basis of the total test score. For each group the percentage of candidates giving each of the responses or omitting the item is shown. The percentages for the whole test population are also given. A good item should have the correct response chosen by the highest percentage in the best group and decreasing percentages by successively poorer groups. On this basis Example 8.4 would be a usable item. The same information can, of course, be shown graphically by means of an *item profile*. Such a treatment, while being clearer to many, does conceal some of the information which is available in the above treatment and is not likely to commend itself to mathematics teachers. An item profile for Example 8.4 is shown in the following diagram.

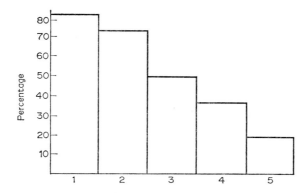

9. *Test Analysis*

When the test has been set the results can be analysed in two ways. Firstly, each of the items can be considered as in the last chapter; from this analysis further information can be obtained to decide whether each item is worth retaining for further re-testing or whether it should be rejected completely. Secondly, the analysis of the test as a whole can be considered.

RELIABILITY

The reliability of an examination is the consistency with which it yields its results. All measurements are subject to certain degrees of error. There are fundamental differences between the statement of a boy's height and his score in a mathematics test. In measuring height there is an agreed standard to which all measurements can be referred; in measuring abilities there is no such agreement. Among other things this means that it is meaningless to talk about *the* reliability of a test. All that can be measured is its reliability when applied to a particular type of pupil who has been trained in mathematics in a particular way. Secondly, in measuring height there is complete agreement about what is being measured—such agreement does not exist in examinations. Thirdly, the act of measuring height does not affect the height being measured—the act of testing may well have an influence on what is being measured. Fourthly, a scale of mathematical ability has neither a true zero nor uniformity of scale. A score of zero on a test of mathematics does not mean that the candidate has no mathematical ability, only that the particular test was not designed in such a way that his ability could be shown; a score of 80 does not represent twice as much ability as a score of 40; indeed such kinds of comparison are meaningless in terms of ability.

However, accepting all these necessary restrictions on the ways in which test scores can be interpreted, it is still desirable that there should be some measurement of the reliability of the score or rank in a test. A candidate's score in a test can be considered as consisting of a 'true' score together with an error term. In symbols $A = T + \varepsilon$. The true score is not to be thought of as some underlying measurement of ability that could be determined if only there were some perfect instrument to measure it with, but rather as the mean value of A if it were possible to apply the test a great number of times to the candidate without there being any change in the candidate between any two testings. The problem in designing an examination is firstly to make ε as small as possible while keeping the test within practical limits, and secondly to have a measurement of ε so that the test results can be meaningfully interpreted. It is useful to consider the sources of this error term ε. There are four main sources which can be listed as follows:

(a) *Marker Unreliability*

Discrepancies between the marking adopted by different examiners or by the same examiner on different occasions and the inability of examiners to agree on the relative merit of a candidate's answer leads to unreliability of marking. The use of fixed-response tests will reduce this error to zero, and it follows that the greater the objectivity of an examination, that is, that any panel of examiners will award the same mark to the same answer, the greater the reliability. Objectivity is a prerequisite to reliability since subjective judgments are inaccurate and therefore unreliable.

(b) *Administrative Unreliability*

Part of the error in measurement in a candidate's score will result from differences in the ways in which a test is administered to different groups of candidates. For example, the time which candidates are given to complete the test may vary from one group to another. While every effort should be made to ensure that the test is a 'power' test rather than a 'speed' test, that is that the test is designed so that the great majority of candidates will have sufficient time to complete the test, it is still true that this aim cannot in practice be completely achieved and a difference of five minutes in the time allowed for the test will affect the scores.

Similarly the scores will be affected by the extent to which different candidates are familiar with the format adopted for the test, and by the instructions which are given to the candidate when the test is being administered. It is the duty of those administering the test to ensure that these circumstances are made as similar as possible for different groups of candidates. The conditions of the test administration must be stated in as much detail as possible, and candidates must be familiarized with the format of the questions before the test. If, for example, the item-matching type is used and some candidates have seen this format before and others have not, the scores will be biased in favour of the former group irrespective of their mathematical ability.

(c) *Test Unreliability*

The test can only be a sample of the mathematics which the candidate is expected to display. It is, however, a structured sample in that the test is constructed to a specification. The reliability of the test is not concerned with the fact that, say, 25 per cent of the marks are awarded for calculus where a different test would give 30 per cent. Such considerations are the province of validity not of reliability. What does concern reliability is the fact that, for example, one test may require the candidate to evaluate the derivative of $\sin (3x + 5)$ whereas a second may require him to evaluate the derivative of $\cos (4x + 7)$; one test may require a knowledge of 7×8 where a second requires 9×6. This kind of sampling of the syllabus will produce errors, or perhaps uncertainties is a better term, in the measured score.

(d) *Candidate Unreliability*

The pupil's physical and mental condition, for example, examination tension, the conditions of testing, motivation and familiarity with the format of the paper are factors associated with the sitting of an examination. There will always be minor changes in the physical and mental state of an examinee from one occasion to another, which will affect his examination performance and consequently the reliability of the examination.

MEASUREMENT OF RELIABILITY

The purpose of measuring reliability is to allow those who are considering the scores on the test to have a knowledge from a candidate's obtained score, of the confidence limits which can be expected for his true score. The kind of information that is wanted is that in a particular test an interval of ± 3 from the obtained score will with 95 per cent confidence contain the true score. The unreliability can therefore be measured by the variance of the 'error' part of the obtained score, but if it is desired to compare different tests for reliability it is more useful to use the ratio of this variance to the total variance of the test. If it is assumed that the error terms are independent of the ability that the test is intended to measure, then the total variance of the test will be the sum of the true variance and the error variance. In symbols,

$$\sigma^2 = \sigma_\infty^2 + \sigma_\varepsilon^2$$

where σ is the total variance, σ_∞ is the true variance
        σ_ε is the error variance.

This equation may be rewritten as

$$1 - \frac{\sigma_\varepsilon^2}{\sigma^2} = \frac{\sigma_\infty^2}{\sigma^2}$$

As has been stated the ratio of the error variance to the total variance is a measure of the unreliability of the test and so it is reasonable to take as a measure of the reliability, the ratio of the true variance to the total variance. This measure is called the coefficient of reliability. (Some writers prefer to use the ratio of the standard deviations rather than of the variances; in this case the ratio is called the index of reliability. One advantage of the latter to the test maker is that since reliabilities are essentially less than one, the index is numerically greater than the coefficient.)

This definition of reliability, while allowing deductions to be made from it on the confidence limits to be assigned to scores, does not directly lead to any means of arriving at a numerical value for it. To arrive at the latter it is necessary to formulate the concept of an equivalent test. Two tests can be said to be equivalent if they are constructed to the same specification, that is they will give the same coverage of syllabus and of ability levels and will have the same order of difficulty. If candidates' scores on two such tests are available, the correlation between these scores is a measure of the reliability of the tests. This result can be easily derived if certain assumptions are made about the nature of the errors which are being measured. It is useful to consider these if only to be aware of the assumptions which are made about the types of error which are being assessed.

The candidate's score on the test can be expressed as $x_t + \varepsilon$ where x_t and ε are characterized by the relationships

$$\Sigma \varepsilon_i = \Sigma \varepsilon_i x_i = 0.$$

In other words the error terms cancel out over the test population as a whole and are independent of the candidate's true score. Thus, to the extent that they are correlated with the candidate's true score, non-mathematical traits in the candidate's behaviour, such as the ability to carry out instructions, will be included in the true score and not in the error term. Any constant error in measurement will also be included in the true score. Since, however, as has already been stated, the scale of measurement has no fixed origin, such constant errors are of no importance. Neither are differences in the scales of measurement in the two tests; if scores on the first test are all twice the scores in the second it will not affect the correlation. It is also assumed that if ε_i and e_i are the error terms in the scores in the two tests, then ε and e are uncorrelated, that is the errors are random errors. With these assumptions it follows simply that the correlation between the two tests measures what has been defined as the reliability of the test.

EQUIVALENT FORMS

The correlation between two equivalent tests provides one measure of the reliability of each test. There are a number of ways in which such equivalent tests can be administered, and each way will give rise to a value of the reliability, all of which may not be the same. One method is to make use of the same test administered on two separate occasions. This—the test, re-test method—is appropriate only when it can be assumed that there will be little or no carry-over from one testing to the other. It may be that candidates will, consciously or not, memorize their responses to the test items. Any such memorization will tend to give a spuriously high value to the measured reliability. On the other hand there may be learning of the particular items between the two test administrations; if this is so the correlation between the scores will be decreased inasmuch as the learning differs between candidates. This type of reliability measurement is probably more appropriate in the assessment of motor skills than in the field of mathematics.

A second possibility is the use of two equivalent forms. These may be administered either in immediate succession or at an appropriate interval. It is desirable that the interval between the two tests should be sufficiently long to allow the differences in candidates' performances due to temporary factors, such as health, to become apparent. Presumably a test attempts to measure something more than a pupil's performance at one particular instant, and this health factor is one of the causes of pupil unreliability which should be assessed. The other factor which it is desired to assess is the effect of slight differences of ability within the syllabus, and so the two tests should be as different as possible while still keeping within the same fairly detailed test specification. However, although this method of measuring reliability is the best, it must be realized that there are great difficulties in trying to implement it, both in the school and in the national situation. In the former case, it will already be realized that the task of preparing an objective test is not an easy one and it is not realistic to imagine that this effort can be doubled. In the national situation no-one could contemplate the increase in examination time which this would require.

SPLIT-HALF RELIABILITY

Where it is not possible to administer two equivalent tests, it is possible to simulate the effect of this by dividing the given test into two halves and treating these as equivalent tests. One of the commoner methods of doing this is to correlate the scores on the even-numbered items with the scores on the odd-numbered items. The assumption in this process is that in the construction of the test, the items will be arranged in a systematic way so that taking every second item will produce a test which will satisfy the specification of the test. When this method is adopted, what is being measured is the reliability of a test which is half as long as the actual test. As would be anticipated, the longer a test the more reliable it will be, assuming of course that the additional items are satisfactory. It is usual to correct the correlation obtained by using the Spearman–Brown formula

$$r_{1+2} = \frac{2r_{12}}{1 + r_{12}}$$

This odd–even method is essentially an administrative convenience and should not be adopted if any other method is possible. It is much more satisfactory to make the conscious effort necessary to split up the test into two halves which will be balanced as far as possible in respect of the abilities they are testing, the difficulty of the items and the coverage of the syllabus.

KUDER–RICHARDSON FORMULA

Any method of determining the reliability of a test from the correlation between two halves necessarily has the disadvantage that an arbitrary decision must be taken on which particular method of splitting the test is to be chosen. There are obviously a great many ways in which a test of forty items may be split into two twenty-item tests, and even when the conditions suggested in the last paragraph have been met there will still remain a number of equally desirable divisions. It is not inconceivable that the reliability obtained may depend on the particular test split chosen. An attempt to overcome this is the use of the Kuder–Richardson Formula 20. This formula yields a reliability estimate similar to that obtained from the previous methods provided the following criteria are satisfied:

(i) The test must be a 'power' test not a 'speed' test. Items which are not attempted by a large number of candidates because they have not been reached will have unreliable facility values and tend to distort the reliability measurement. It is reasonable to assume that this is so, since a test which does not, in fact, allow candidates to make any response is bound to have an in-built overmeasurement of consistency. With experience it is possible to know the time necessary for a particular form of test and so this is not likely to cause much difficulty.

(ii) The test must be reasonably homogeneous. The derivation of the formula involves the assumption that the average value of the covariance of items within the test is equal to the average value of the covariance of items between the test and a parallel form. Unless the test is homogeneous this assumption will not be satisfied and the value obtained will underestimate the true reliability. In most cases, however, this underestimation will not be serious. The formula for calculating the reliability is

$$r_{11} = \frac{n}{n - 1}\left[\frac{s^2 - \Sigma\, p_i q_i}{s^2}\right]$$

where n is the number of items in the test,
  s is the standard deviation of the test,
  p_i is the proportion of candidates responding correctly to item i,
  $q_i = 1 - p_i$.

VALIDITY

The second attribute of a test which an attempt is made to measure is its validity, that is the extent to which it measures what it sets out to measure. It should be noted that a test which is not reliable cannot be valid, but a reliable test need not be a valid one. Thus the measurement of candidate's height would be a reliable measure of their mathematical ability but not a valid measure.

In considering the question of validity, the first question which must be posed is, 'valid for what?' 'Valid as a test of mathematical ability' does not answer this question. It is necessary to question further and ask whether the test is meant to be valid as a test of:

(a) the knowledge which the candidate has acquired, or
(b) the desired attributes which the course was intended to inculcate, or
(c) the candidate's fitness to proceed further in mathematics.

In most cases, as has been seen, the test is forced into attempting a number of different roles.

Measurements of validity are essentially measurements of correlation. The candidate's score on the test is correlated with his criterion score in the attribute which the test attempts to measure. One difficulty in this procedure is that the criterion score will not usually be a perfectly reliable measure. In order to compensate for this unreliability it is necessary to have two measures of the criterion score, A and B say. If r_{1A} denotes the correlation between the test and the criterion score A, and r_{1B} and r_{AB} have similar meanings, then it can be shown that the correlation between the test and the 'true' criterion score is

$$\frac{r_{1A}r_{1B}}{r_{AB}}$$

Since there are a number of purposes which most examinations are intended to measure, there are a number of different aspects of validity. Only in some of these cases is it possible to put a numerical value on the validity. Different aspects of validity are given different names, and various writers may not use the same name for the same aspect. Some of the different aspects of validity are content, predictive and concurrent.

(a) *Content Validity*

This refers to the extent that the examination relates to the instructional objectives of the course and whether or not it is a representative sample of the total syllabus it is supposed to be testing. There are thus two aspects of this content, or face, validity. If a test is composed of items, 90 per cent of which test knowledge, and the objectives of the course emphasize higher level abilities, then the test cannot be valid. Such a test would be said to lack objective validity. On the other hand, a test may have its items correctly distributed over the various ability levels—correctly here meaning in accordance with the aims of the syllabus maker—but not cover the syllabus adequately. Thus a test in which all the items were concerned with calculus would not be valid—it would lack content validity. Such a test might be reliable and might even correlate very highly with future performance in mathematics but it would not have content validity.

Content validity cannot be measured directly. The best way to ensure this type of validity is to construct the test so that it conforms to a test specification which has been laid down in advance by those who are responsible for the syllabus. It must be emphasized that the time to make up this test specification is not when the test is being constructed but when the syllabus is being planned. The syllabus planners must ask and answer such questions as 'What are the objectives of the course?', 'What are the relative weights to be given to the various parts of the syllabus?', 'How much of the course is intended to produce knowledge learning only?'. If a course is planned with these questions in mind, it will be relatively easy to construct a test to assess them properly. Too often in the past the process has been reversed and the examination has determined the objectives of the course.

(b) *Predictive Validity*

This refers to the ability of the test to predict future behaviour. This future behaviour may be short term, as when a teacher uses a school examination to predict success in a national examination, or longer term, as when results in an examination at school level are used to select, say, for university entrance and therefore for success in a university course. It is quite possible for a test to possess this type of validity while lacking content validity completely. Thus a mathematics test might well have predictive validity for a physics course higher than that of a school physics examination but it would be completely lacking in content validity. It may be argued that in a sense predictive validity is the only type of validity that matters, not in the narrow sense that the test measures the ability to go on successfully with mathematics at tertiary level, or to achieve success in some other field, but in a much wider sense.

All teaching is intended to produce change in the person taught, and this change can only be worthwhile in so far as it is permanent. A test will therefore be valid if it measures these permanent changes which the course is intended to produce. It is thus predicting future behaviour. The difficulty in practice is, of course, to obtain some measure of these changes. In some cases the difficulty may only be one of the time interval between the test and what it is intended to predict, and in this case it may be possible to make use of some intermediate criteria. Thus in using a test to predict success in a university course, there will be a gap of four or five years before the test can be assessed. If, however, it is assumed that success in the first year of such a course is a measure of final success, then the first year scores can be used as an intermediate criterion. The result must, however, be treated with caution. A high value will only imply some degree of validity, a low value may not imply the absence of validity in that the intermediate criterion may not measure some elements which are common to the test and the ultimate criterion.

(c) *Concurrent Validity*

This refers to the correspondence between performance on the test and performance on another form of test which is known to be valid. The principle use of this concept would be where a traditional type of examination was available, known to be valid, and it was desired to replace this by a different form for purely administrative reasons. Very seldom, however, is this the case. When such replacements are made, they are usually motivated as much by a desire to improve the measurement as for other reasons. In such a case a very high validity is not to be expected nor, in fact, to be desired.

INCREASING RELIABILITY AND VALIDITY

Measurements such as have been described are made to investigate whether the testing instrument is suitable for use. If unacceptable values are obtained, the test maker must take steps to raise them. The validity of a test can only be increased by carefully analysing the purpose of the test and seeking to match the form more closely to the purpose. The reliability can be increased by increasing the number of items in the test provided that the additional items conform to the test specification. It is possible to show that if a test is increased so that there are n times as many items as in the original test, the reliability will be increased from its original value r to

$$\frac{nr}{1 + (n - 1)r}$$

Thus to increase the reliability of a 40-item test from 0.8 to 0.9 would require an additional 50 items. Spectacular increases in reliability cannot be expected. It is also possible to increase the reliability by increasing the number of distractors in each item, since this will decrease the probability of chance scores, but again there is a limit to the extent to which this can be done.

TEST CONSTRUCTION

Points of possible help in designing an efficient examination are as follows:

(i) There is an obvious relationship between the length of a test and its reliability since an increase in the number of items increases the reliability, but there is no point in increasing the length in order to secure this higher reliability at the expense of validity and usability. If questions are added then they must be typical of those already in the test. The work 'inadequacy' is often used to indicate the degree to which a test is of sufficient length to sample widely the achievement of the objectives it is constructed to measure.

(ii) Research at present indicates that the more detailed and specific the steps in an examination procedure, the more reliable and valid the end-product tends to be.

(iii) For maximum efficiency in achieving its objective of arranging candidates in rank order, the examination must discriminate among the examinees with the highest possible degree of reliability and validity. An examination cannot be valid unless it is also reliable and discriminating, where the discriminating power of the examination is its ability to distinguish between candidates who have ability in mathematics and those who have not.

(iv) Validity is specific to purpose, content, objectives and pupil level, and is a matter of degree. For example, a mathematics aptitude examination may have a high degree of validity for predicting success in university courses, an adequate validity for indicating application of mathematical principles and a low validity for indicating the ability to recall factual knowledge.

(v) In general, achievement or attainment examinations should possess high content validity, and aptitude, selection and prediction examinations high predictive validity.

(vi) High validity is the main characteristic that a test should possess, even if attaining it means a slight loss in reliability and any of the other characteristics, for example objectivity, for without validity the test is useless regardless of how high the degree of the rest of the characteristics can be made.

Examination boards are resorting more and more to the science of testing, and teachers should acquire some familiarity with the techniques and their use in order to bring to their own examinations a greater degree of reliability and validity.

10. *Assessment*

It cannot be too strongly emphasized that all assessment of human ability or achievement is approximate. Unlike physical measurement, the degree of probable error is fairly large. Not only the measuring instruments but also the individual cannot be expected to produce consistent results. To interpret examinations or test marks in the same light as the measurement of height or weight will lead to disappointment and frustration.

The first step in assessing ability or attainment is to determine the purpose of the assessment. The use of one measuring instrument may yield an unsatisfactory result. For example, a written examination in mathematics may give a result which does not indicate a pupil's skill in the use of mathematical instruments. Once the purpose of the assessment is known, the nature of the measuring instruments can be determined.

GENERAL ASSESSMENT

Assessments are made in general by the following measuring instruments:

1. *Oral Questioning*

Good classroom teaching is generally accompanied by regular questioning from which the teacher forms a rough assessment of the pupil's potential ability for a subject and his level of training. Such an assessment is highly subjective and unreliable, although over a period a teacher can gain an impression of the extent to which pupils have mastered the subject. Oral examinations are time-consuming and cannot readily be used for large numbers of candidates in external examinations.

2. *Assignments*

A written description of a project is widely used to assess work which has been carried out over a lengthy period. Marks awarded for work of this nature are highly subjective but the technique ranks high as an accepted method of individual assessment in higher education. One of the advantages of this method is that it can cover both practical and theoretical work.

3. *Traditional Written Examinations*

Questions in this type of examination require the pupils to respond in writing and in their own words or to supply the solution to a problem with the constructed answer varying in length from a few lines to several pages. The expression 'free response' is often used to describe a traditional solution since the pupil is allowed freedom of response in answering the question, there often being no single answer to it which is accepted as correct and complete. The candidate has to recall information and formulate his own answers, and hence the testing situation is unstructured and is of particular importance in testing, for example, the ability to create and develop logical proof.

Even when the mark scheme is detailed, the marks in this type of examination tend to be subjective. Other disadvantages include a limited sampling of knowledge and abilities which results in relatively low reliability and hence low validity. Due to the small number of questions that can be attempted by the candidates in any one examination confined to traditional-type questions, only relatively few areas of the syllabus can be sampled and there is a danger that the selection of questions is not representative of the major objectives of the course. Obviously this type of item can provide areas for examining in depth but it can also lead to 'spotting' of certain popular recurring areas by the pupil, making success or failure in the examination a function of chance. If he is lucky then the questions selected for the examination may deal with aspects he knows well, although he may be lacking in his knowledge of the subject, but on the other hand he may be heavily penalized if he knows relatively little about the majority of the chosen questions. It is difficult also to estimate the comparability and reliability of a particular set of traditional questions since a pupil, due to the wide range of choice of questions, can choose a combination not attempted by any other candidate. For example, the choice of six out of ten is typical in an examination, and one candidate could attempt questions 1, 2, 3, 4, 5, 6, another 6, 7, 8, 9, 10. Both may obtain passes and appear in the same order of merit, but the two candidates have been assessed on what is in effect different examinations. It is also very difficult to set ten questions of comparable standard.

4. Structured Questions

This type of question lies somewhere between the traditional type and the objective type. It takes the form of a relatively complicated situation broken down into a series of related questions, each of which requires the pupil to supply a short answer. An essential feature of this item, which must be taken into account in construction, is that of clarity rather than conciseness.

Example 10.1
For a school concert, x people each bought a 15p ticket, whilst the rest of those who went to the concert, y people, each bought a 25p ticket. The total amount of money received from the sale of those tickets was £126.

(a) Write down the number of pence in £126.
(b) Write down, in terms of x, the number of pence received from the sale of the 15p tickets.
(c) Write down, in terms of y, the number of pence received from the sale of the 25p tickets.
(d) Use the results of (b) and (c) to write down, in terms of x and y, the total number of pence received from the sale of all tickets.
(e) Use the results of (a) and (d) to form an equation in x and y.
(f) Write down a second equation, in x and y, given that the total number of tickets sold was 600.
(g) Solve your two equations to find the values of x and y.
(h) If each of the 15p tickets had cost 20p and each of the 25p tickets had cost 30p, what would have been the total money received from the sale of all tickets at those new prices?

Example 10.2

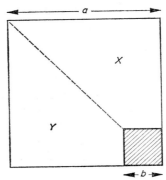

The diagram represents a square piece of cardboard of side *a* units. From it is removed the small shaded square of side *b* units, and the remainder is cut along the dotted line to form the two exactly identical shapes *X* and *Y*.

(a) Write down in terms of *a* and *b*,
   (i) the area of the original square of cardboard,
   (ii) the shaded area removed
   (iii) the total area of *X* and *Y*.
(b) Make a sketch showing how *X* and *Y* may be combined to form a rectangle.
(c) Write down in terms of *a* and *b*,
   (i) the length of the longest side of the rectangle,
   (ii) the length of the shortest side of the rectangle,
   (iii) the area of the rectangle, using the lengths of the sides just written down.
(d) By considering your answers to (a) (iii) and (c) (iii) state an algebraic relationship illustrated in this example.

5. *Practical Examinations*

An important element in mathematics is the ability and skill to handle and use accurately instruments such as the ruler, set square, protractor, compasses, and the more sophisticated machines such as the calculating machine and the computer. In the classroom it is possible to roughly assess a pupil's ability and attainment, and even practical examinations are possible. In external examinations pupils are allowed to use either slide rules or calculating machines as alternatives to logarithmic tables.

6. Objective Tests

Objective tests consist of questions framed in such a way as to give only one predetermined correct answer. Accordingly the marking of the questions is objective—although subjectivity is still involved in the setting of the questions.

The advantages of objective tests are:

(i) This type of examination makes it possible to attain wide syllabus coverage. The large number of items included in an examination of this form, 50 to 100 items, can be attempted in the time taken to answer 4 or 5 traditional questions, makes a broad sampling of the content and specific objectives possible, and so reduces the unreliability and increases the validity of the examination. Each item can be chosen to cover a specific point of the syllabus under test. The complete range of instructional objectives can therefore be tested eliminating the 'spotting' tendency of candidates which is present in traditional type examinations.

(ii) Rapidity of marking is ensured in an objective test since there is only one correct response to each item, and skilled markers are not required for purposes of correction. Examination marking machines have been developed and their use will eventually become widespread. If machines are not available then the scoring can be delegated to clerical staff.

The speed with which an objective test can be marked is also helped by the normal procedure of awarding one mark to each question regardless of difficulty. This is possible for two reasons: first the difficulty of each item is determined before the candidates sit the test, and second the candidates have to attempt all the questions and are therefore examined on the whole syllabus. This practical advantage of the objective test over the traditional test increases with the number of pupils to be tested and is of particular value in the case of the external examination and possibly when the number of candidates in a school examination is greater than 100.

(iii) One of the major strengths of the objective examination is the objectivity of the marking process. Errors arising from discrepancies in marking are reduced to negligible proportions, but note that subjectivity is involved in the setting of the examination, since the items are selected on the subjective judgment of the examiners. Thus while marking is perfectly objective and reliable, the complete examining procedure is not so.

(iv) The method that is generally adopted in the construction of an objective test, namely via a specification in which a definite weighting can be assigned to each syllabus topic and ability category by a proportionate increase in the number of items devoted to it, ensures that the subjectivity that enters into the setting of any examination, for example one set of examiners could choose a different set of items for the same course, is considerably reduced.

The two-way specification also ensures not only a balanced paper and the elimination of the unreliability of inadequate sampling of instructional objectives, but also high content validity. There is also evidence to indicate that well-constructed objective tests can have high reliability and acceptable concurrent validity.

(v) An objective test can test successfully the attainment of not only the abilities of knowledge and memorization of facts, for which it is ideally suited, but also those of application of principles and the higher abilities.

(vi) If an examination is to be acceptable then it must possess the characteristics described in Chapter 9 but it follows that the items, if they are to contribute to the effectiveness of the examination, must each possess the same characteristics. In objective testing the process of pre-testing, that is trying the test out in advance, is necessary to achieve this.

It enables the difficulty of each item to be obtained before the pupils for whom the test is designed sit it, and ensures control of the level of difficulty of all the items. This is important in the case of an achievement test, for example, where the reliability can be improved by ensuring that the items are neither too difficult nor too easy, since low reliability can result from a restricted spread of scores.

The pre-test also enables not only the test but also each item to discriminate between pupils who have ability in mathematics and those who have not. From the pre-test, each item is provided with a discriminating factor, and if it is not discriminating satisfactorily it is rejected. Hence it is possible to produce an examination of the correct length with all the items discriminating in the correct direction and of the correct order of difficulty for the particular candidate population. This results in an increase in the reliability of the examination.

Pre-testing, then, is a major advantage that the objective test has over the traditional test, since in the case of the latter it is usual to analyse the results after the test and adjust the marks accordingly, which is highly unsatisfactory and unreliable.

(vii) Due to the highly structured nature of the objective test, each item can be constructed to test one particular objective in one particular content area, that is it asks one precise question to which there is one unique answer, and hence this type of testing is extremely useful in diagnosing the difficulties encountered by individual pupils in the classroom, as well as indicating to the teacher the success or otherwise of the learning experiences. It also means that a pupil cannot bluff his way through a question, as he is forced to respond to items designed to elicit certain behaviours without the need for ability and skill in writing.

(viii) In an objective examination the candidate indicates each answer by encircling a letter or placing a tick in a box, on a special sheet of paper, enabling him to spend the major part of the examination thinking, whereas a traditional examination is wasteful of time for thought since one quarter to one third of the time allocation is spent in writing.

The limitations of objective examinations are:

(i) The construction of objective type papers is a skilled and time-consuming task. To overcome this, the skilled markers who are now available are being instructed, by experts on item construction, in the techniques of designing this type of examination. Gradually panels of trained item writers are being built up in the subjects which are now using this form of assessment. What is more important is that the panels are made up of teachers who can return to their schools and apply in the classroom the techniques that they have acquired.

It is also recognized that the time spent by a teacher in preparing an objective test is worthwhile in terms of forcing him to think seriously about the instructional objectives of the course and the learning experiences he devises to enable his pupils to attain them.

(ii) There are other factors that can affect a pupil's score in an objective test. For example, the correct answer can be obtained by guessing or by badly constructed items, thus reducing the reliability of the test. However, it is true to say that random guessing takes place much less than is supposed. Note that a different type of guessing occurs in traditional examinations, with the teacher and pupils attempting to predict the questions which might be asked. Increasing the number of responses for each item improves reliability and hence, although true–false questions are somewhat unreliable in multiple-choice questions the problem of guessing, although not eliminated, is diminished.

Another factor which can occur, if care is not taken in construction, is when the item is attempting to test the attainment of a higher ability. In many cases, by necessity, the question is long and success may depend on the reading ability of the pupil, which is obviously not the ability being tested if it is a mathematics examination. If too many such items are included in the examination, a resultant distortion in the marks can result which reduces the overall validity.

(iii) A backwash factor must result, to some extent, from all examinations, and in particular tends to develop when a different form of examination question is introduced. Care must be taken to ensure that the examination always reflects and encourages teaching methods, is designed to measure the predetermined objectives of the course, and is not constructed to test that which is easily examinable. Coaching of, and practice with, pupils in this type of testing will take place, which is natural in order to familiarize them with the form of the examination, but efficient and intensive coaching is by no means synonymous with good education and, although it will give no advantage in tackling easier questions, it may be helpful in coping with more difficult and complex items. There might also be a lack of attention paid to the setting down of solutions to problems and the writing of continuous English if those abilities were never examined directly.

Regardless of the type of examination, it is essential to ensure that the effects of backwash be positive and useful to the pupils. For example, in writing questions related to the objectives of a course, aspects of them which had not previously been clear may be revealed, enabling next year's pupils to benefit from an improved list.

(iv) It is difficult to test successfully the attainment of creative ability, fluency of expression, the ability to select and organize ideas, and the ability to synthesize and develop a logical argument, by objective methods of examining, and such skills are better tested by other types of assessment.

Although it is generally accepted that objective items can be constructed to test the higher abilities, in practice this depends upon the skill and ingenuity of the item writer. Consequently, until teachers have had experience in the construction and use of such tests, traditional questions will probably provide for some time the most convenient method of assessing attainment of the complex skills, especially if the class is small or when time for preparation of the examination is limited.

Objective tests are superior to traditional tests in overall reliability and validity. However, it must not be assumed that objective testing is appropriate for assessing all the declared objectives of a course, and present work in this field would seem to indicate that when a written examination, external or internal, is necessary, then objective and traditional questioning both have a part to play. Each type is efficient for some purposes and inefficient for others. For specificity, comprehensiveness and the testing of factual knowledge over a broad area of subject matter and of pupils whose essay writing ability is restricted, objective-type items are superior in order to take advantage of their wider sampling coverage and greater reliability; whereas if the need is to test more complex abilities, such as creative ability and the ability to arrange and evaluate material and ideas in an ordered way, then despite their limitations the traditional type question is probably better. An objective paper to cover the breadth of a syllabus, that is to ensure extensive sampling, and a traditional paper to test the depth, that is to ensure intensive sampling, would appear to be the best combination at present, but note that it is not advisable to mix the two types in one paper.

7. Teachers' Assessment

One of the most reliable forms of assessment of ability and attainment is the teachers' assessment. When a pupil has been studying mathematics for a few years in a school, a continuous record of his progress can be kept and a careful study of this gives a reasonably accurate assessment of the pupil's mastery of the subject. One of the weaknesses, however, of teachers' assessment is that the standards still vary from school to school and from teacher to teacher. In order to make such assessments comparable, they can be scaled so that they are made to have the same mean, scatter and distribution.

ASSESSMENT IN MATHEMATICS

At the end of a four- or five-year course, the most reasonable and logical method of obtaining a final assessment of a pupil's achievement in mathematics would be to base it on three factors:

(a) the results of an external traditional examination,
(b) the results of an external objective examination,
(c) the teacher's assessment.

The third factor would be based on internal examinations, class work both written and observed, and possibly some type of attitude assessment with the necessary imposition of scaling in order to obtain a uniform standard for all schools.

Bibliography

BLOOM, B. S. *Taxonomy of Educational Objectives, Handbook I—Cognitive Domain* (London: Longmans, 1956), *Handbook II—Affective Domain* (London: Longmans, 1964)

DAVE, R. H. *Paper in Developments in Educational Testing, Volume I* (University of London Press Ltd., 1969)

DAVIS, F. B. *Item Analysis Data* (Harvard Educational Papers Number 2)

FURST, E. J. *Constructing Evaluation Instruments* (New York: McKay, 1964)

GRONLUND, N. E. *Measurement and Evaluation in Teaching* (New York: Macmillan, 1965)

HUSÉN, T. (Ed.) *International Study of Achievement in Mathematics: A comparison of twelve countries* (New York: Wiley, 1967 2 volumes)

MAGER, R. F. *Preparing Instructional Objectives* (Fearon Publications, 1962)

MORSE, H. T. and McCUNE, G. H. *Selected Items for the Testing of Study Skills and Critical Thinking* (Washington, D.C.: National Council for the Social Studies, 1964)

PHILIPPS, R. C. *Evaluation in Education* (Columbus, Ohio: Merrill, 1968)

Scottish Education Department Consultative Committee on the Curriculum, *Curriculum Paper 7—Science for General Education* (H.M.S.O., 1969)

Further Reading

AVITAL, S. M. and SHETTLEWORTH, S. J. *Objectives for Mathematics Learning* (Ontario Institute for Studies in Education)

BLOOM, B. S., HASTINGS, J. T. and MADAUS, G. F. *Handbook on Formative and Summative Evaluation of Student Learning* (New York: McGraw-Hill, 1971)

BROWN, J. *Objective Tests: Their Construction and Analysis* (London: Longmans, 1966)

Examinations and Assessment: Mathematics Teaching Pamphlet No. 14 (Association of Teachers of Mathematics)

H.M.S.O. Examinations Bulletins of the Schools Council.

No. 2 The Certificate of Secondary Education: *Experimental Examinations: Mathematics I*, 1964

No. 3. The Certificate of Secondary Education: *An introduction to some techniques of examining*, 1964

No. 4. The Certificate of Secondary Education: *An introduction to objective-type examinations*, 1964

No. 8. The Certificate of Secondary Education: *Experimental Examinations: Mathematics I*, 1965

McINTOSH, D. M. *Statistics for the Teacher* (Oxford: Pergamon, 1967)

McINTOSH, D. M., WALKER, D. A. and MACKAY, D. *The Scaling of Teachers' Marks and Estimates* (Edinburgh: Oliver and Boyd, 1962)

NUTTALL, D. L. and SKURNIK, L. S. *Examination and Item Analysis Manual* (National Foundation for Educational Research in England and Wales 1969)

Scottish Certificate of Education Examination Board, *Objective Testing; Mathematics, Candidates' Booklet* (Gibson)

Scottish Certificate of Education Examination Board, *Objective Testing; Mathematics, Teachers' Booklet* (Gibson)

WISEMAN, S. and PIDGEON, D. A. *Curriculum Evaluation* (National Foundation for Educational Research in England and Wales 1970)